Temperature and Plant Development

Temperature and Plant Development

Edited by

KEARA A. FRANKLIN

PHILIP A. WIGGE

WILEY Blackwell

Editorial Offices
1606 Golden Aspen Drive, Suites 103 and 104, Ames, Iowa 50010, USA
The Atrium, Southern Gate, Chichester, West Sussex, PO19 8SQ, UK
9600 Garsington Road, Oxford, OX4 2DQ, UK

For details of our global editorial offices, for customer services and for information about how to apply for permission to reuse the copyright material in this book please see our website at www.wiley.com/wiley-blackwell.

Library of Congress Cataloging-in-Publication Data

Temperature and plant development / edited by Keara A. Franklin, Philip A. Wigge.
pages cm
Includes bibliographical references and index.
ISBN 978-1-118-30820-2 (cloth)
1. Plants–Effect of temperature on. 2. Plants–Development. I. Franklin, Keara A. II. Wigge, Philip A.
QK755.T455 2014
581.3–dc23

2013033686

A catalogue record for this book is available from the British Library.

Wiley also publishes its books in a variety of electronic formats. Some content that appears in print may not be available in electronic books.

Cover design by Matt Kuhns.

Set in 10.5/12pt TimesNewRoman by SPi Publisher Services, Pondicherry, India
Printed and bound in Malaysia by Vivar Printing Sdn Bhd

1 2014

Contents

Color plate section is located between pages 130 and 131.

Contributors

Bob Baxter	School of Biological and Biomedical Sciences, Durham University, Durham, UK
Ralph Bours	Laboratory of Plant Physiology, Wageningen University, Wageningen, The Netherlands
Andreas W. Ebert	AVRDC – The World Vegetable Center, Shanhua, Tainan, Taiwan
Kathleen Greenham	Department of Biological Sciences, Dartmouth College, Hanover, NH, USA
Karen J. Halliday	School of Biological Sciences, The University of Edinburgh, Edinburgh, UK
Jerry L. Hatfield	USDA-ARS National Laboratory for Agriculture and the Environment, Ames, IA, USA
Jian Hua	Department of Plant Biology, Cornell University, Ithaca, NY, USA
Dong-Hwan Kim	Section of Molecular Cell and Developmental Biology, Institute for Cellular and Molecular Biology, the University of Texas at Austin, Austin, TX, USA
Dana MacGregor	Biosciences, University of Exeter, Exeter, UK
C. Robertson McClung	Department of Biological Sciences, Dartmouth College, Hanover, NH, USA
Harriet G. McWatters	School of Biological Sciences, The University of Edinburgh, Edinburgh, UK
Garry J. O'Leary	Department of Primary Industries, Horsham, Victoria, Australia

Steven Penfield Biosciences, University of Exeter, Exeter, UK

Thijs L. Pons Plant Ecophysiology, Institute of Environmental Biology, Utrecht University, Utrecht, The Netherlands

P. V. Vara Prasad Department of Agronomy, Kansas State University, Manhattan, KS, USA

Marcel C. G. Proveniers Molecular Plant Physiology, Institute of Environmental Biology, Utrecht University, Utrecht, The Netherlands

Robert J. Redden Department of Primary Industries, Horsham, Victoria, Australia

Sibum Sung Section of Molecular Cell and Developmental Biology, Institute for Cellular and Molecular Biology, the University of Texas at Austin, Austin, TX, USA

Gabriela Toledo-Ortiz School of Biological Sciences, The University of Edinburgh, , Edinburgh, UK

Shyam S. Yadav Ministry of Agriculture, Irrigation & Livestock, Government of Islamic Republic of Afghanistan, Kabul, Afghanistan

Martijn van Zanten Molecular Plant Physiology, Institute of Environmental Biology, Utrecht University, Utrecht, The Netherlands

Preface

Temperature is a key environmental signal regulating plant growth and development. Small changes in ambient temperature can affect a wide range of processes throughout the plant lifecycle, from seed germination and plant architecture through to flowering and reproductive development. Prolonged low-temperature treatment can act as a seasonal cue, signaling the onset of winter to prime flowering and seedling development the following spring. In addition to providing important environmental information, exposure to temperature extremes can adversely affect plant survival. The evolution of developmental adaptations to withstand prolonged cold or heat has enabled some species to exploit ecological niches in adverse habitats. In more temperate regions, many plants have evolved acclimation responses to minimize cellular damage associated with freezing and heat stress.

Plants can detect temperature changes as small as 1°C. Despite the importance of temperature in controlling plant growth and survival, our current understanding of how temperature signals are perceived is rudimentary. Suggested thermosensory mechanisms include changes in membrane fluidity, activation of membrane transport channels, altered protein activity, and the direct regulation of gene expression through altered DNA accessibility. Molecular dissection of plant temperature responses has revealed significant crosstalk with light and circadian signaling pathways. The integration of temperature and photoperiod signals provides plants with accurate seasonal information, priming adaptive responses to adverse conditions while preventing the wasteful allocation of resources in milder climates. In natural environments, plants are subject to multiple environmental signals simultaneously, resulting in trade-offs between different stress responses. It is perhaps, therefore, not surprising that temperature has significant effects on plant immunity and defense signaling.

Climate change presents major challenges for global agriculture and the preservation of ecosystems and biodiversity. Current climate models predict future increases in global temperature, with potentially devastating effects on crop production. Relatively moderate increases in ambient temperature (<6°C) can invoke dramatic changes in plant development, reducing harvest yield. Increased temperature additionally accelerates flowering, promoting floral development out of synchronization with pollinator lifecycles. Indirect impacts of elevated temperature on plant development include increased pathogen susceptibility and enhanced water use. The latter may promote leaf cooling in well-watered environments but would have severe consequences in drought conditions. Conversely, exposure to sudden frost can lead to catastrophic crop losses in nonacclimated species. Understanding how plants perceive, integrate, and respond to temperature signals may provide novel molecular targets for the production of crops resilient to climate change and inform predictions as to the impact of global warming on plant ecology and biodiversity. Enhanced knowledge of plant temperature responses could additionally lead to more energy-efficient horticultural production. The stature and flowering time of glasshouse crops are commonly controlled through manipulations of light and temperature. Understanding how plants perceive and respond to small temperature changes at different times of day in diverse light environments will greatly facilitate the design of optimal growth management regimes.

This volume is designed to provide a comprehensive and up-to-date account of the role of temperature in plant development. It is aimed at all students and teachers of modern plant biology, academics with an interest in the environmental regulation of development, and policy makers working in the area of climate change, ecology, and global food security.

Keara A. Franklin,
University of Bristol, UK
Philip A. Wigge,
Sainsbury Laboratory,
University of Cambridge, UK

1 Temperature sensing in plants

Steven Penfield and Dana MacGregor

1.1 Introduction

Plants are subjected to considerable variations in temperature, both daily and annually, and are surprisingly temperature-sensitive organisms: it has been shown that levels of cytosolic calcium in plant cells can respond to as little as a 1°C temperature shift, and a 4°C diurnal temperature cycle is sufficient to entrain the circadian clock (Knight and Knight 2000; McClung et al. 2002). Developmental processes such as seed germination can be completely inhibited by 1–2°C temperature rises (Argyris et al. 2011). Despite the experimentally described responsiveness of plant physiology and development to temperature, no thermosensory molecule has yet been unequivocally identified.

1.2 Passive and active temperature responses in plants

Since the nineteenth century, it has been suggested that temperature affects the rates of biological reactions according to the thermodynamic principles that govern chemical reactions more generally. In this scenario, the rate of reactions whose activation energy is significantly greater than a given temperature will increase proportionally to the exponent of the temperature rise. As a guiding principle, the free energy change normally dictates that for biological reactions in biologically-relevant temperature ranges, rates will increase roughly twofold to threefold with a 10°C rise in temperature. This figure, known as the temperature coefficient or Q_{10}, became popular during the twentieth century following its popularization after the work of Van't Hoft (1896) and Arrhenius (1889). Thus, specific biological processes, from the growth of bacteria to the respiration of plants, were shown to have a Q_{10} of approximately 2–3 within a range often described as 'room temperature' (Běhrádek 1930). This law has even been applied to developmental biology and plant growth, a good example of which is the study of seed germination

Temperature and Plant Development, First Edition. Edited by Keara A. Franklin and Philip A. Wigge.

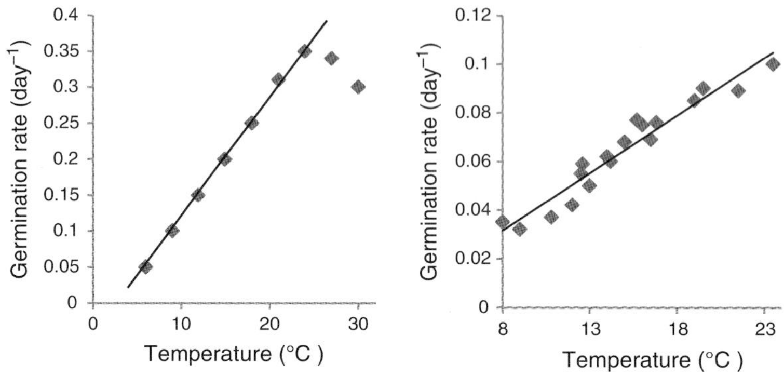

Figure 1.1 Speed of germination in seeds often shows a relationship indicative of a passive temperature response. In the lab (left) and in the field (right), carrot seeds germinate at a rate with a linear relationship with temperature, with a *Q*10 of around 2. Redrawn from Hegarty (1973).

by Hegarty (1973). Here it was shown that the speed of germination of many common vegetable seeds showed a dependency on temperature consistent with that expected by the passive effects of temperature in biochemical reactions (Figure 1.1). In a complex biological event such as seed germination, requiring respiration, cell division, and cell elongation, it was postulated that temperature affects the rate of a single, vital rate-limiting reaction in the process. Where Q_{10}s were found to differ dramatically from 2 to 3, it was suggested that this might be due to a complex effect of temperature on many reactions with unpredictable consequences (Běhrádek 1930).

Many types of biomolecule are expected to be subject to these types of 'passive' temperature effects including the fluidity of lipid bilayer membranes, the conformations of proteins, and the behavior of nucleic acids. Clearly, organisms need to be able to control their physiology to maintain performance of vital functions over a range of possible biochemical reaction rates. However, among this sea of events that must continue to function equivalently at multiple temperatures, one or more are used as temperature sensors by biological organisms. Here we will advance a broad definition of a temperature sensor as a *passive temperature-controlled change in configuration of a molecule or assembly of molecules that is integrated with downstream signal transduction in order to create an active signal regulating a process of adaptive significance*. Such signaling pathways are referred to as 'active' temperature responses because during signal transduction, the effects can be amplified or buffered such that the temperature coefficient may differ significantly from 2 to 3. Plants use temperature information to allow them to synchronize their life -history with the seasons or to adapt their physiology to different temperature environments.

1.3 Temperature sensing during transcriptional regulation

Steady-state (SS) transcript levels of many plant genes are highly sensitive to temperature, and there are several documented cases where these changes are essential for known adaptive responses. In seeds, a 10°C temperature variation causes changes in approximately 10-fold more expressed genes than in seedlings (Kendall et al. 2011), showing that different plant tissues have different innate sensitivities of transcription rates to temperature. In prokaryotes, thermodynamic effects of temperature affect gene expression through the control of chromosomal supercoiling and by conformational changes in topoisomerases that control the supercoiling process. In thermophilic bacteria, the enzyme reverse gyrase acts to induce tight supercoiling even at extreme temperatures (>80°C), and the action of this enzyme appears to keep the DNA context in a configuration that permits a reasonable speed of transcription (Forterre et al. 1996). Therefore, key to understanding the temperature-control of transcription in eukaryotes will be an analysis of how DNA interacts with its environment to control transcription rate and how this varies over temperature.

A well-studied example of the importance of temperature signaling in development is vernalization, the requirement for a prolonged cold period before plants are able to respond to floral-inductive signals. In *Arabidopsis*, the vernalization pathway acts through downregulation of the floral repressor *FLOWERING LOCUS C* (*FLC*) (reviewed in Amasino, 2010, and discussed further in Chapter 4, Sheldon et al. 1999; Michaels and Amasino 1999). In vernalization-requiring accessions, stable repression of *FLC* is achieved after several weeks of cold exposure. Here we will focus on the primary steps in responding to temperature at the *FLC* locus, of which there are two key points. The first of these is that after 2 weeks of cold, the accumulation of the repressive mark trimethylated histone 3 lysine residue 7 (H3K27) begins at the *FLC* locus. This begins to appear around the transcription start site and its presence correlates with the downregulation of *FLC* expression (Bastow et al. 2004). A second repressive mark, dimethylation of H3K9, also appears upstream of the transcription start site, and both marks are necessary for the maintenance of *FLC* repression upon transfer to cold. H3K27me3 modifications require the activity of the Polycomb Repression Complex 2 (PRC2), and mutations of several subunits of these have been identified in forward genetic screens for *Arabidopsis* mutants unable to induce or maintain the vernalized state. These mutants share a common phenotype in that the repression of *FLC* is initiated in the cold, but is not maintained upon transfer to the warm (Gendall et al. 2001; Sung and Amasino 2004; Greb et al. 2007). One of these, *VERNALIZATION INSENSITIVE3* (*VIN3*), is itself upregulated at the transcript level by prolonged low-temperature exposure, suggesting that the temperatureregulation of *VIN3* expression is a primary event during vernalization. However, although VIN3 is necessary for the stable repression of *FLC* after vernalization, expression of *VIN3* alone cannot confer a vernalization-like response on unvernalized plants (Sung and Amasino 2004), showing that control of VIN3 and by extension PRC2 complex abundance is unlikely to be the primary temperature signal for vernalization. How temperature controls *VIN3* expression is not known.

Recently it has been suggested that the observed quantitative repression of *FLC* by increasing durations of cold reflects a bistable switch in *FLC* repression occurring in ever greater numbers of cells during vernalization (Angel et al. 2011). In support of this, non-saturating vernalization exposures result in a cell autonomous response in which some cells are silenced for *FLC:GUS* expression, while many cells continue to express at high levels. After 2 weeks of cold, H3K27me3 levels begin to increase around the transcription start site of *FLC* full-length transcript, suggesting this is a primary response to cold in the vernalization pathway (Angel et al. 2011). The dynamics of this increase correlate well with the timing of the increase in *VIN3* expression, suggesting that the VIN3 protein might play a role in the targeting of the PRC2 complex to the H3K27me3 nucleation site. However, the lack of any predicted or known sequence specificity of VIN3 for any DNA sequence and the inability of *VIN3* overexpression to induce a vernalization response suggest that H3K27me3 modification in response to VIN3 cannot alone explain the temperature responsiveness of *FLC* transcript SS levels.

The *FLC* locus also produces at least two noncoding RNAs that appear to have a role in the vernalization process. The first, designated *COOLAIR*, is a long antisense transcript that covers the entire *FLC* locus and has a promoter that can independently confer cold responsiveness to a reporter gene independently of gene context (Swiezewski et al. 2009). This latter observation appears to tie temperature responsiveness to transcriptional control, rather than RNA stability. Importantly, *COOLAIR* expression occurs in *vin3* mutants and was also shown to confer downregulation of the sense *FLC* transcript, suggesting that coldresponsiveness is VIN3 independent. However, *COOLAIR* is unlikely to be solely responsible for the downregulation of sense *FLC* expression in *Arabidopsis*, since T-DNA insertion mutants lacking the *COOLAIR* transcript but expressing a functional FLC protein continue to show a robust vernalization response (Helliwell et al. 2011). More recently, a second noncoding but this time sense transcript-designated *COLDAIR* has been identified with a role in the vernalization response (Heo and Sung 2011). The *COLDAIR* transcript is transcribed from the first intron, a region of *FLC* long known to have a role in the control of vernalization, and has as part of its promoter an approximately 300 bp sequence known as the vernalization response element (VRE; Sung et al. 2006). *COLDAIR* knockdown lines show reduced repression of *FLC* by vernalization. It is therefore suggested that the *COLDAIR* transcript controls H3K27me3 nucleation at the *FLC* coding transcription start site (Heo and Sung 2011). All this activity occurs long before *VIN3* expression increases, suggesting that activity surrounding the transcriptional promotion of *COLDAIR* is a key primary step in responding to temperature during vernalization. A key question for future vernalization response remains understanding the mechanism of how temperature regulates *FLC* expression.

In wheat, a key player in the vernalization response is the MADS-box transcription factor VERNALIZATION1 (VRN1, see Trevaskis et al. 2007). Instead of cold conferring transcriptional repression, expression of VRN1 is activated during vernalization to promote flowering. Repression of *VRN1* expression also requires the first intron (Fu et al. 2005; Cockram et al. 2007), but cold activation appears to be primarily driven by elements in the *VRN1* promoter. So, in wheat, the first intron is again required for repression, but not for cold activation. The similarities and differences between *Arabidopsis FLC* and wheat *VRN1* regulation are striking and highlight how much of our understanding of these processes reflects the transcriptional regulatory processes downstream of the temperature-sensing pathways and much less is known of how temperature signals are sensed. In addition, it is still completely unknown how vernalization integrates temperature signaling with time in order to measure the duration of the cold signal during winter.

The complex kinetics of the control of *FLC* expression has led to the search for alternative models for studying the control of transcription by temperature. Several plant genes seem to have quantitative responses of SS mRNA levels to environmental temperature over a wide temperature range. Good examples of these in *Arabidopsis* are *HEAT SHOCK PROTEIN 70* (*HSP70*) and *COLD-REGULATED 15a* (*COR15a*), which increase transcript abundance in response to increasing and decreasing temperatures, respectively (Penfield 2008; Kumar and Wigge 2010). A genetic screen for *HSP70* mis-regulation has been used to identify genes necessary for correct temperature responses in the ambient temperature range (between about 12°C and 27°C) (Kumar and Wigge 2010). The first mutants characterized were novel alleles of *ACTIN-RELATED PROTEIN6* (*ARP6*), encoding a component of the plant SWR1 complex required for the deposition of the histone 2A variant H2A.Z into chromatin (Mizuguchi et al. 2004). ARP6 is necessary for coordinating the temperature transcriptome, since the warm-temperature transcriptome is constitutively expressed at lower temperatures in *arp6* alleles, lacking H2A.Z incorporation. Consistent with this observation, H2A.Z nucleosomes are evicted from chromatin at higher temperatures, enabling RNA Pol II to transcript genes such as *HSP70* that are induced at higher temperatures (Kumar and Wigge 2010).

An interesting feature of H2A.Z biology is that the effect of H2A.Z nucleosomes on transcription appears to be locus specific, since, for example, in the case of *FLC*, H2A.Z deposition is correlated with transcription of the gene (Deal et al. 2005), while in the case of *HSP70*, and genes involved in the phosphate starvation response, H2A.Z loss results in upregulation of expression (Smith et al. 2010). When H2A.Z occupancy is analyzed by chromatin immunoprecipitation (ChIP), higher temperatures result in a decrease in occupancy (Kumar and Wigge 2010), and this effect appears to be independent of the transcriptional status of the gene, suggesting that eviction of H2A.Z occurs in response to warmer temperature, independent of effects on transcription (Franklin 2010; Kumar and Wigge 2010). For SS mRNA levels to rise with increasing temperature, transcription rates must increase at a rate exceeding that of mRNA degradation rates. If it is assumed that mRNA degradation rates follow a simple Arrhenius–Van't Hoft relationship with temperature, additional mechanisms in addition to changes in nucleosome occupancy may be required to cause a gain in SS transcript levels. An increase in SS levels may require a positive feedback loop, either induction of an activator of transcription (Figure 1.2) or a factor which stabilizes mRNA. Larger effects of increasing

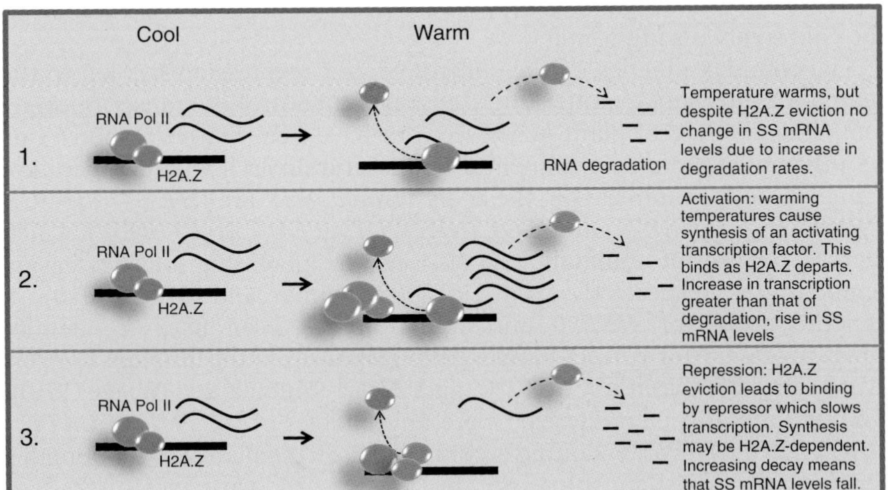

Figure 1.2 Control of SS transcript abundance by the temperature-dependent association of H2A.Z-containing nucleosomes with transcription start sites. (1) A passive affect alone does not produce an increase in SS levels, because mRNA degradation rates too are affected. (2) The eviction of H2A.Z allows an activator to bind. The abundance of this may also be temperaturecontrolled or a positive feedback through autoactivation. (3) Eviction allows a repressor to bind, allowing nucleosome eviction to depress transcription rates. After Kumar and Wigge (2010) and Franklin (2010). RNA Pol II– RNA polymerase II. For color detail, please see color plate section.

temperature on decay rates may act to lower SS RNA levels even in the presence of increasing transcription. It seems likely that H2A.Z nucleosome eviction is likely tied to other processes in order to drive temperature-regulated changes in SS mRNA levels.

While H2A.Z nucleosomes are required for the normal behavior of the ambient temperature transcriptome, it has not been demonstrated directly that they are themselves thermosensors. To fulfil our definition of a temperature sensor, we must also show that our sensor is linked by a signal transduction pathway to processes of adaptive significance. Have plants exploited the temperature responsiveness of the association of DNA and H2A.Z nucleosomes to confer thermosensitivity to important developmental or physiological processes? As we are primarily concerned with temperature regulation of transcription, the imperative is to discover temperature- and *arp6*-dependent changes in gene expression that underlie plant adaptation to variable environments. *arp6* mutants do indeed show phenotypes that suggest they are compromised in their ability to regulate important temperature-controlled plant processes, such as the timing of flowering and hypocotyl elongation (Deal et al. 2005; Kumar and Wigge 2010). In *arp6*, flowering is earlier than wild type in long days and short days. If H2A.Z nucleosomes are rate limiting for the expression of key flowering regulators, such as *FT*, then higher temperatures could cause flowering by triggering H2A.Z eviction. Supporting this hypothesis, the transcription factor PIF4 directly activates *FT* in a temperature-dependent manner, and the ability of PIF4 to bind to the *FT* promoter is temperature dependent (Kumar et al. 2012). However, in short days shifting *arp6* plants from 22°C to 27°C results in a large decrease in flowering time, demonstrating that a substantial portion of the thermosensory response is ARP6 independent. This suggests that other components alongside SWR1 and H2A.Z have an important role in thermosensation controlling flowering time. Taken together, these results suggest that chromatin accessibility appears to be temperature regulated, but whether this is a direct response to temperature or controlled by a temperature-regulated chromatin modifying pathway remains to be seen.

Warm temperatures increase *Arabidopsis* hypocotyl elongation in an auxin-dependent manner (Gray et al. 1998). This process requires the activity of the PIF4 transcription factor (Koini et al. 2009), and the binding of PIF4 to auxin biosynthesis genes has been shown to be temperature regulated, supporting a role for temperature-mediated changes in chromatin accessibility in controlling this process (Franklin et al. 2011). Consistent with this, *arp6* have elongated hypocotyls (Kumar and Wigge 2010). In a close parallel to the role of ARP6 in the control of flowering time, *arp6* hypocotyls do elongate in response to temperature but show a reduced response. This appears to reinforce the idea that this pathway must function redundantly with others to control ambient temperature responses in plants.

1.4 Sensing cold: A role for plasma membrane calcium channels in plants

In animals, it is now well established that voltage-dependent action transporters in the TRP family of potassium channels are necessary for sensing temperature (Peier et al. 2002). Plants do not contain conserved relatives of these proteins, suggesting that the mechanism for temperature sensing is not shared. However, electrophysiological evidence suggests that plant cell membranes depolarize with a decrease in temperature and that this depolarization is accompanied by an influx of calcium in to the cytosol (Minorsky 1989; Knight et al. 1991, 1996). The degree of depolarization and increase in cytosolic calcium is dependent not only on the degree of cooling but also on the rate (Plieth et al. 1999). This type of response has parallels with the downstream transcriptional activation of the C-repeat binding factor (CBF) family, involved in plant cold acclimation (discussed in Chapter 2). For instance, when plants are shifted from a growth temperature of 20–10°C, the corresponding increase in *CBF* transcript levels is less than that resulting from the larger change in temperature from 20°C to 4°C (Zarka et al. 2003). For *Arabidopsis* grown at 20°C, a shift to 14°C or lower is sufficient to induce *CBF* transcripts to detectable levels, and plants acclimated to growth at 4°C will induce a *CBF* response when shifted below 0°C (Zarka et al. 2003). Interestingly, daily calcium oscillations also modulate the responsiveness of cytosolic free calcium concentrations ($[Ca^{2+}]^{cyt}$) to cold, providing a potential mechanism for circadian gating (Dodd et al. 2006). Calcium influx after cold appears to cause the depolarization rather than be a response to it, because calcium channel blockers such as lanthanum also block cold-induced membrane depolarization (Lewis and Spalding 1998). This shows that extracellular calcium influx is a primary plant response to cold and cooling. Despite many efforts to identify the types of channel responding to the cold stimulus, none have been found, possibly indicating a high degree of genetic redundancy. Given that these responses occur within a minute or so of a cold pulse, it is likely that these are early signaling events in the detection of a cold stimulus. Ion influx is an attractive system for generating an active temperature signal since work has previously been done to establish the resting membrane potential, enabling a large response to be achieved by simply opening the channel, exploiting the resting potential to amplify the temperature signal. The next question to answer is whether these cold-induced oscillations in $[Ca^{2+}]^{cyt}$ led to a downstream signal transduction cascade.

Efforts to determine transcriptional responses to elevated $[Ca^{2+}]^{cyt}$ have helped to identify an ongoing cold signal transduction influencing gene expression (summarized in Figure 1.3). Experiments which have involved the pharmacological manipulation of $[Ca^{2+}]^{cyt}$ showed that elevations

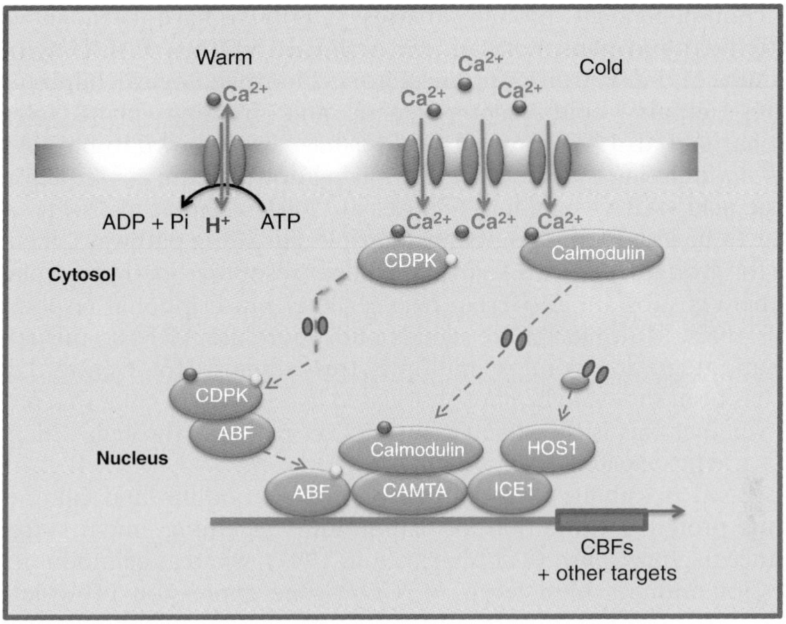

Figure 1.3 Schematic of our current understanding of cold perception and signaling in plants, showing possible signal transduction pathways from cytosolic free calcium increases to transcriptional control. For abbreviations see main text. For color detail, please see color plate section.

in $[Ca^{2+}]^{cyt}$ led to the activation of transcription of genes with ABA RESPONSE ELEMENTS (ABREs) in their promoters (Galon et al. 2010). If a LUCIFERASE reporter gene driven by ABREs is transformed into tobacco, luciferase activity displays $[Ca^{2+}]^{cyt}$ sensitivity. This analysis was extended by Whalley et al. (2011) who developed an elegant system in which $[Ca^{2+}]^{cyt}$ oscillations were induced by applying voltages to seedlings floating in cuvettes. Here it was found that genes containing several types of cis-elements in their promoters were induced by $[Ca^{2+}]^{cyt}$, including ABREs, CAMTA-binding elements, C-repeats, and TCP-binding sites. Many of the induced genes have previously been shown to be cold responsive (Zarka et al. 2003). Together this work shows convincingly that changes in $[Ca^{2+}]^{cyt}$ can be transduced as a signal to promoters and affect gene expression. Carpaneto et al. (2007) showed that mutants deficient in cold signaling components were not deficient in the control of $[Ca^{2+}]^{cyt}$ oscillations, suggesting that $[Ca^{2+}]^{cyt}$ oscillations are truly upstream of all known signal transduction. This leaves us with the problem of elucidating the elements of the signal transduction pathway between cold-induced calcium oscillations and gene expression.

Calcium-dependent protein kinases (CDPKs) have been shown to directly bind and phosphorylate the ABRE-BINDING FACTOR (ABF) subfamily of bZIP transcription factors. Pharmacological inhibition of CDPKs impairs cold responsiveness and freezing plant tolerance (Tähtiharju et al. 1997). Several CDPKs have been shown to bind ABF1–ABF4, modulating their activity and having downstream consequences for abscisic acid (ABA) signaling (Zhu et al. 2007; Zhao et al. 2011). ABFs appear to be integrating points for multiple signalling pathways, including those for drought, salt, ABA, and cold sugar responses, perhaps explaining why there is substantial overlap between the transcriptional responses to these signals. Multiple abiotic signals phosphorylate ABFs so this is likely to be an integration point of multiple stress signals. However, to date, no ABF has been shown to be phosphorylated in response to cold, so it remains open whether this is the signal transduction pathway through which cold signals are propagated.

A second possibility is a central role for calmodulin and calmodulin-binding proteins. Inhibition of calmodulin signaling impairs freezing tolerance in *Arabidopsis* (Tähtiharju et al. 1997), whereas calmodulin over-expression induces high levels of *COR* gene expression (Townley and Knight 2002). A dissection of the *CBF2* promoter revealed a conserved binding element of the calmodulin-binding CAMTA family of transcription factors which was able to confer cold responsiveness to a reporter gene (Doherty et al. 2009). Of the six CAMTA transcription factors in *Arabidopsis*, CAMTA3 could be shown to bind the conserved *CBF2* promoter element, whereas deletion of either *CAMTA1* or *CAMTA3* both impaired the acquisition of freezing tolerance. This work clearly demonstrated a role for CAMTAs in cold signal transduction. An interesting feature of this study is that the authors report that loss of CAMTAs also impairs the ability of the *CBF2* promoter to respond to calcium signals that result from mechanical stimulation (Knight et al. 1991). This suggests that the calmodulin signal transduction pathway is not specific for cold but can potentially carry signals from multiple stresses to downstream target promoters (Doherty et al. 2009), some of which also stimulate increases in $[Ca^{2+}]^{cyt}$. Thus, it is again possible that cold signaling uses a pathway that is shared with other processes and may not itself exist as a separable entity.

A final class of calcium signal transduction proteins worthy of consideration as transducers of a cold-induced calcium influx are the calcineurin B-like proteins (CBLs; Luan et al. 2002). CBLs are small calcium-binding proteins which function through the binding of a large family of serine/threonine protein kinases known as CBL-INTERACTING PROTEIN KINASES (CIPKs). This pathway also appears to have overlapping functions in the response to multiple stresses and in ABA signal transduction. At least one member, CIPK3, is induced transcriptionally by cold and is required for cold-responsive gene expression (Kim et al. 2003). CBLs and CIPKs may therefore be part of a positive feedback pathway that amplifies cold signals in plants.

1.5 A role for membrane fluidity as an upstream temperature sensor?

The control of membrane fluidity is evident across organisms from mammals to microorganisms. In plants, manipulation of fatty acid content can impair ability to survive temperature extremes, including chilling and heat stress (Miquel et al. 1993; Kodama et al. 1995; Murakami et al. 2000). In keeping with the hypothesis that homeostasis of membrane fluidity is required for normal cellular functions and stress resistance, depleting unsaturated fatty acids, for instance, through *fad2* mutation, exposes plants to chilling injury, whereas overexpression of *FAD7* improves chilling tolerance (Miquel et al. 1993; Kodama et al. 1995). Homeostasis of membrane fluidity is achieved by the regulation of the activity of fatty acid desaturases, at either the protein or the transcriptional level (Gibson et al. 1994; Matsuda et al. 2005). Interest in the possibility that membrane fluidity could act as a temperature sensor in plants comes primarily from the study of prokaryotic systems, in which a signal transduction cascade is initiated by changes in fluidity that result in small changes in membrane thickness (see Penfield et al. 2012). In this system, a protein with a dual role as phosphatase and kinase switches from a phosphatase to a kinase as membrane thickness increases with decreasing temperature (Cybulski et al. 2010). This in turn activates a phosphorelay that increases transcription of a fatty acid desaturase, whose products act to restore membrane fluidity to its original level.

Homologues of these proteins are not conserved in plants, and the plant fluidity sensor remains to be discovered. However, in homeostasis lays the major problem with this system as a general temperature sensor: if fluidity is constant across multiple temperatures, how can fluidity act as an absolute temperature sensor? Conclusive evidence that changes in membrane fluidity regulate processes other than membrane composition are currently lacking in any experimental system, from prokaryotes to eukaryotes. In plants, the main evidence comes from the study of the application of chemicals which artificially alter membrane fluidity such as dimethylsulfoxide (DMSO, a rigidifier) and benzyl alcohol, a liquefier. Application of these compounds leads to alteration in temperature-controlled transcription (Orvar et al. 2000; Sangwan et al. 2001). However, genetic manipulation of membrane fluidity does not result in similar changes, leading to speculation that the effects of DMSO and BA were not mediated by their effect on fluidity. Measuring fluidity is not straightforward, but recently an elegant assay using photoquenching of membrane-bound GFP allowed direct measurements of fluidity in plants for the first time (Martiniere et al. 2011). This study showed that membrane fluidity is not perfectly homeostatic in plants, but subject to diurnal variation. This variation is only incompletely counterbalanced by daily changes in composition, probably due to the slow turnover rates of membrane lipids. This work too could not find a convincing link between fluidity changes and the control of adaptive temperature responsiveness, suggesting that the major role of temperature sensing by membrane fluidity changes may be to control the downstream activity of fatty acid desaturases and maintain membrane fluidity homeostasis.

1.6 Temperature sensing by proteins

So far we have considered the roles of cation transporters and direct transcriptional responses to temperature and how these have been co-opted for the regulation of adaptive processes in plants. There is also ample evidence that transcription factors are themselves subject to a post-translational regulation by temperature, although it is not always clear where the temperature signal is perceived. A few examples are given as follows.

One of the best-characterized pathways that directly links a change in temperature to a protein level response is the control of INDUCER OF COLD EXPRESSION 1 (ICE1) protein levels, an MYC-type basic helix-loop-helix transcription factor, by HIGH EXPRESSION OF OSMOTICALLY RESPONSIVE GENES 1 (HOS1). In this process, the level of *CBF* transcript is determined by the concentration of the transcriptional activator ICE1, whose levels are reduced by cold. *HOS1* encodes an E3 ubiquitin ligase that is involved in targeting ICE1 for degradation by the proteasome (Ishitani et al. 1998; Lee et al. 2001; Dong et al. 2006). When the plant is experiencing a warm temperature, GFP-tagged HOS1 is located in the cytoplasm and its transcript levels are constant (Lee et al. 2001). Within 10 min of a shift to cold, *HOS1* transcript levels drop dramatically, allowing for a brief alleviation of ICE1 ubiquitination, and because polyubiquitination targets proteins for degradation by the proteasome, the levels of ICE1 then increase (Dong et al. 2006). The mechanism that results in the reduction in *HOS1* transcript level in response to a drop in temperature is unidentified, but it is short lived. After being in the cold for 1 h, *HOS1* transcript levels return to pre-cold levels, and GFP-tagged HOS1 is found preferentially in the nucleus. With HOS1 levels rising in the nucleus, it is able to interact with and ubiquitinate ICE1, targeting it to the proteasome for degradation so that *CBF3* and *CBF*-targeted transcripts return to low levels (Dong et al. 2006). Thus, the temperature-controlled accumulation of HOS1 in either the nucleus or cytoplasm is proposed to be critical in the way in which the ICE1 protein transiently activates CBF transcription after cold shock.

A second related process where temperature-regulated transcription factors control adaptive processes in plants is the temperature control of growth. Low temperatures inhibit plant growth, not just through a passive effect on reaction rates, but through an active process of the regulation of plant hormone levels, notably gibberellins and auxin (Penfield 2008). Increases in auxin levels that accompany warm temperatures require PHYTOCHROME-INTERACTING FACTOR 4 (PIF4), a protein whose DNA-binding activity is altered with increasing temperatures (discussed in Chapter 3; Franklin et al. 2011; Kumar et al. 2012). It is currently not clear whether PIF4 itself is affected by temperature, for instance, through a post-translational modification or complexing with inhibitory factors, or whether PIF4 is competing with another factor, such as histone binding (Kumar and Wigge 2010). In this context, it is useful to consider two further PIFS whose abundance is directly affected by temperature, SPATULA (SPT) and LONG HYPOCOTYL IN FAR-RED1 (HFR1). SPT controls temperature-responsive seed dormancy and growth (Penfield et al. 2005; Sidaway-Lee et al. 2010), with mutants having high growth rates even at low temperatures and overexpressors showing a dwarf phenotype only if the temperature is low. Importantly, cold elevates SPT protein levels leading to growth repression, in a manner that closely parallels the function of DELLA proteins. In contrast, HFR1 protein levels and *HFR1* transcription are increased at warm temperatures (Foreman et al. 2011). It appears that temperaturecontrol of multiple PIFs and their activities underlie plant growth responses to temperature, although the mechanism by which temperature affects PIF protein levels and activity remains unknown.

Circadian clocks are also temperature sensitive and must maintain a constant 24-hour period across a broad range of temperatures (discussed in Chapter 6). This defining feature of circadian oscillators is especially interesting and requires a mechanism to buffer period against the effect of temperature on individual rates, such as transcription and mRNA decay, termed temperature compensation. For instance, transcription and mRNA decay rates, on average, increase by 3.5-fold for a 10°C temperature rise in plants (Sidaway-Lee and Penfield, unpublished data). The *Arabidopsis* circadian clock is composed of a series of interlocking negative feedback loops with dawn- and dusk-expressed components. The genes that make up the core clock components include a pair of highly homologous MYB transcription factors called CIRCADIAN CLOCK-ASSOCIATED 1 (CCA1) and LATE ELONGATED HYPOCOTYL (LHY) which function synergistically and have partially redundant functions in the repression of dusk-expressed TIMING OF CAB EXPRESSION1 (TOC1), ELF3, ELF4, AND LUX ARRYTHMO (LUX; see McWatters and Devlin, 2011, for a recent review of our understanding of clock architectures in plants). Genetic studies have identified various clock genes without which temperature compensation fails, including LHY, PRR9, and GI (Edwards et al. 2006; Gould et al. 2006; Salome et al. 2010). However, a recent systems biology approach produced the first temperature-compensated model of the *Arabidopsis* circadian clock and predicted that there is one major temperature-sensitive clock protein and that is LHY (Gould et al. 2013). As temperature increases, LHY protein levels increase and this in turn increases the time after dawn at which the transcript levels of TOC1 and the other evening expressed genes, all inhibited by LHY, start to rise (Gould et al. 2013). This increase in LHY protein levels slows the clock and therefore balances faster rates in other reactions and allows the clock to maintain a 24-hour period at warmer temperatures. How this rise in LHY protein levels is mediated and whether the temperature sensitivity of LHY can be exploited for downstream signaling is currently unknown.

Several components of the LHY/CCA1 to PRR regulatory loop have additionally been tied to control of *CBF* transcript accumulation. Plants that are arrhythmic in constant conditions due to the loss of *PSEUDO-RESPONSE REGULATOR 9* (*PRR9*), *PSEUDO-RESPONSE REGULATOR 7* (*PRR7*), and *PSEUDO-RESPONSE REGULATOR 5* (*PRR5*) have constitutively elevated levels of *CBF* transcripts and increased survival when challenged with freezing, suggesting that PRR9, PRR7, and PRR5 act as direct negative regulators of *CBF* transcript levels (Nakamichi et al. 2009). Because LHY/CCA1 and these PRR proteins are locked in a translational feedback loop, it is possible that the PRR proteins are acting indirectly through LHY and CCA1. Support for this hypothesis has recently been published in a paper by Dong et al. (2011) in which the authors present ChIP experiments demonstrating direct binding of GFP-tagged CCA1 to the *CBF1–CBF3* promoters. Their evidence indicates that LHY also plays a role in regulating *CBF* transcript levels as single *lhy* or *cca1* mutants have nearly wild-type levels and rhythms of *CBF2* and *CBF3* (Dong et al. 2011). Experiments with *lhy cca1* double mutants show that *LHY* and *CCA1* are necessary for the rhythmic accumulation of *CBF* and *CBF*-target gene transcripts under constant temperature and for their induction in response to a challenge with cold (Dong et al. 2011). Therefore, it is now clear that *CBF* transcript levels are rhythmically regulated by the activity of LHY and CCA1 and that their induction in response to cold requires these MYB transcription factors, which are themselves temperature regulated.

1.7 Summary

The evidence presented so far suggests that there are likely to be several temperature sensors in plants, all of which have roles to play in adaptive plant responses to varying environmental temperatures. A multitude of unknowns also suggests that as a field, we have a long way to go before we can say we understand how plants sense temperature. Further research into how protein levels are temperature regulated and how temperature-controlled transcription controls plant development will undoubtedly lead to further important insights into the way in which plants cope with living in seasonal environments and suggest mechanisms through which changes in environmental temperature might act to drive evolution.

References

Amasino R (2010) Seasonal and developmental timing of flowering. Plant J. 61: 1001–1013.

Angel A, Song J, Dean C, Howard M (2011) A polycomb-based switch underlying quantitative epigenetic memory. Nature 476: 105–108.

Argyris J, Truco MJ, Ochoa O, McHale L, Dahal P, Van Deynze A, Michelmore RW, Bradford KJ (2011) A gene encoding an abscisic acid biosynthetic enzyme (LsNCED4) collocates with the high temperature germination locus Htg6.1 in lettuce (Lactuca sp.). Theor. Appl. Genet. 122: 95–108.

Arrhenius S (1889) Uber die Reaktionsgeschwindigkeit bei der Inversion von Rohrzucker durch Sauren. Zeitschrift fuer Physikalische Chemie 4: 226–248.

Bastow R, Mylne JS, Lister C, Lippman Z, Martienssen RA, Dean C (2004) Vernalization requires epigenetic silencing of FLC by histone methylation. Nature 427: 164–167.

Běhrádek, J (1930) Temperature coefficients in biology. Biol. Rev. 5: 30–58.

Carpaneto A, Ivashikina N, Levchenko V, Krol E, Jeworutzki E, Zhu JK, Hedrich R (2007) Cold transiently activates calcium-permeable channels in *Arabidopsis* mesophyll cells. Plant Physiol. 143: 487–494.

Cockram J, Chiapparino E, Taylor SA, Stamati K, Donini P (2007) Haplotype analysis of vernalization loci in European barley germplasm reveals novel VRN-H1 alleles and a predominant winter VRN-H1/VRN-H2 multi-locus haplotype. Theor. Appl. Genet. 115: 993–1001.

Cybulski LE, Martín M, Mansilla MC, Fernández A, de Mendoza D (2010) Membrane thickness cue for cold sensing in a bacterium. Curr. Biol. 20: 1539–1544.

Deal RB, Topp CN, McKinney EC, Meagher RB (2005) Repression of flowering in *Arabidopsis* requires activation of FLOWERING LOCUS C expression by the histone variant H2A.Z. Plant Cell 19: 74–83.

Dodd AN, Jakobsen MK, Baker AJ, Telzerow A, Hou SW, Laplaze L, Barrot L, Poethig RS, Haseloff J, Webb AA (2006) Time of day modulates low-temperature Ca signals in *Arabidopsis*. Plant J. 48: 962–973.

Doherty CJ, Van Buskirk HA, Myers SJ, Thomashow MF (2009) Roles for *Arabidopsis* CAMTA transcription factors in cold-regulated gene expression and freezing tolerance. Plant Cell 21: 972–984.

Dong CH, Agarwal M, Zhang Y, Xie Q, Zhu JK (2006) The negative regulator of plant cold responses, HOS1, is a RING E3 ligase that mediates the ubiquitination and degradation of ICE1. Proc. Natl. Acad. Sci. USA 103: 8281–8286.

Dong MA, Farré EM, Thomashow MF (2011) Circadian clock-associated 1 and late elongated hypocotyl regulate expression of the C-repeat binding factor (CBF) pathway in *Arabidopsis*. Proc. Natl. Acad. Sci. USA 108: 7241–7246.

Edwards KD, Anderson PE, Hall A, Salathia NS, Locke JC, Lynn JR, Straume M, Smith JQ, Millar AJ (2006) FLOWERING LOCUS C mediates natural variation in the high-temperature response of the *Arabidopsis* circadian clock. Plant Cell 18: 639–650.

Foreman J, Johansson H, Hornitschek P, Josse EM, Fankhauser C, Halliday KJ (2011) Light receptor action is critical for maintaining plant biomass at warm ambient temperatures. Plant J. 65: 441–452.

Forterre P, Bergerat A, Lopezgarcia P (1996) The unique DNA topology and DNA topoisomerases of hyperthermophilic Archaea. FEMS Microbiol. Rev.18: 237–248.

Fu DL, Szucs P, Yan LL, Helguera M, Skinner JS (2005) Large deletions within the first intron in VRN-1 are associated with spring growth habit in barley and wheat. Mol. Genet. Genomics 273: 54–65.

Franklin KA (2010) Plant chromatin feels the heat. Cell 140: 26–28.

Franklin KA, Lee SH, Patel D, Kumar SV, Spartz AK, Gu C, Ye S, Yu P, Breen G, Cohen JD, Wigge PA, Gray WM (2011) Phytochrome-interacting factor 4 (PIF4) regulates auxin biosynthesis at high temperature. Proc. Natl. Acad. Sci. USA 108: 20231–20235.

Galon Y, Finkler A, Fromm H (2010) Calcium-regulated transcription in plants. Mol. Plant 3: 653–669.

Gendall AR, Levy YY, Wilson A, Dean C (2001) The VERNALIZATION 2 gene mediates the epigenetic regulation of vernalization in *Arabidopsis*. Cell 107: 525–535.

Gibson S, Arondel V, Iba K, Somerville C (1994) Cloning of a temperature-regulated gene encoding a chloroplast omega-3 desaturase from *Arabidopsis* thaliana. Plant Physiol. 106: 1615–1621.

Gould PD, Locke JC, Larue C, Southern MM, Davis SJ, Hanano S, Moyle R, Milich R, Putterill J, Millar AJ, Hall A (2006) The molecular basis of temperature compensation in the *Arabidopsis* circadian clock. Plant Cell 18: 1177–1187.

Gould PJ, Ugarte N, Domijan M, Costa M, Foreman J, Macgregor D, Rose K, Finkenstädt B, Penfield S, Rand DA, Millar AJ, Halliday KJ, Hall AJW (2013) Network balance via cryptochrome signalling controls the *Arabidopsis* circadian clock over ambient temperatures. Mol. Syst. Biol. 9, 650.

Gray WM, Ostin A, Sandberg G, Romano CP, Estelle M (1998) High temperature promotes auxin-mediated hypocotyl elongation in *Arabidopsis*. Proc. Natl. Acad. Sci. USA 95: 7197–7202.

Greb T, Mylne JS, Crevillen P, Geraldo N, An H, Gendall AR, Dean C (2007) The PHD finger protein VRN5 functions in the epigenetic silencing of *Arabidopsis* FLC. Curr. Biol. 17: 73–78.

Hegarty TW (1973) Temperature coefficient (*Q*10), seed germination and other biological processes. Nature 243: 305–306.

Helliwell CA, Robertson M, Finnegan EJ, Buzas DM, Dennis ES (2011) Vernalization-repression of *Arabidopsis* FLC requires promoter sequences but not antisense transcripts. PLoS One 6: e21513.

Heo JB, Sung S (2011) Vernalization-mediated epigenetic silencing by a long intronic noncoding RNA. Science 331: 76–79.

Ishitani M, Xiong L, Lee H, Stevenson B, Zhu JK (1998) HOS1, a genetic locus involved in cold-responsive gene expression in *Arabidopsis*. Plant Cell 10: 1151–1161.

Kendall SL, Hellwege A, Marriot P, Whalley C, Graham IA, Penfield S (2011) Induction of dormancy in *Arabidopsis* summer annuals requires parallel regulation of DOG1 and hormone metabolism by low temperature and CBF transcription factors. Plant Cell 23:2568–2580.

Kim KN, Cheong YH, Grant JJ, Pandey GK, Luan S (2003) CIPK3, a calcium sensor-associated protein kinase that regulates abscisic acid and cold signal transduction in *Arabidopsis* Plant Cell 15: 411–423.

Knight MR, Campbell AK, Smith SM, Trewavas AJ (1991) Transgenic plant aequorin reports the effects of touch and cold-shock and elicitors on cytoplasmic calcium. Nature 352: 524–526.

Knight H, Trewavas AJ, Knight MR (1996) Cold calcium signaling in *Arabidopsis* involves two cellular pools and a change in calcium signature after acclimation. Plant Cell 8: 489–503.

Knight H, Knight MR (2000) Imaging spatial and cellular characteristics of low temperature calcium signature after cold acclimation in *Arabidopsis*. J. Exp. Bot. 51: 1679–1686.

Kodama H, Horiguchi G, Nishiuchi T, Nishimura M, Iba K (1995) Fatty acid desaturation during chilling acclimatization is one of the factors involved in conferring low-temperature tolerance to young tobacco leaves. Plant Physiol. 107: 1177–1185.

Koini MA, Alvey L, Allen T, Tilley C, Harberd N, Whitelam GC, Franklin KA (2009) High temperature-mediated adaptations in plant architecture require the transcription factor, PIF4. Curr. Biol. 19: 408–413.

Kumar SV, Wigge PA (2010) H2A.Z-containing nucleosomes mediate the thermosensory response in *Arabidopsis*. Cell 140: 136–147.

Kumar SV, Lucyshyn D, Jaeger KE, Alós E, Alvey E, Harberd NP, Wigge PA (2012) Transcription factor PIF4 controls the thermosensory activation of flowering. Nature 484: 242–245.

Lee H, Xiong L, Gong Z, Ishitani M, Stevenson B, Zhu JK (2001) The *Arabidopsis* HOS1 gene negatively regulates cold signal transduction and encodes a RING finger protein that displays cold-regulated nucleo-cytoplasmic partitioning. Genes Dev. 15: 912–924.

Lewis BD, Spalding EP (1998) Nonselective block by La3+ of *Arabidopsis* ion channels involved in signal transduction. J. Membr. Biol. 162: 81–90.

Luan S, Kudla J, Rodriguez-Concepcion M, Yalovsky S, Gruissem W (2002) Calmodulins and calcineurin B-like proteins: Calcium sensors for specific signal response coupling in plants. Plant Cell 14 (suppl.): S389–S400.

Martinière A, Shvedunova M, Thomson AJ, Evans NH, Penfield S, Runions J, McWatters HG (2011) Homeostasis of plasma membrane viscosity in fluctuating temperatures. New Phytol. 192: 328–337.

Matsuda O, Sakamoto H, Hashimoto T, Iba K (2005) A temperature-sensitive mechanism that regulates post-translational stability of a plastidial omega-3 fatty acid desaturase (FAD8) in *Arabidopsis* leaf tissues. J. Biol. Chem. 280: 3597–3604.

McClung CR, Salomé PA, Michael TP (2002) The *Arabidopsis* circadian system. In: Somerville CR, Meyerowitz EM (eds). The *Arabidopsis* book. American Society of Plant Biologists, Rockville, MD. http://www.aspb.org/publications/arabidopsis/ (accessed on August 20, 2013).

McWatters HG, Devlin PF (2011) Timing in plants – a rhythmic arrangement. FEBS Lett. 585: 1474–1478

Michaels SD, Amasino RM (1999) FLOWERING LOCUS C encodes a novel MADS domain protein that acts as a repressor of flowering. Plant Cell 11: 949–956.

Minorsky P (1989) Temperature sensing by plants: a review and hypothesis. Plant Cell Environ.12: 119–135.

Miquel M, James D Jr, Dooner H, Browse J (1993) *Arabidopsis* requires polyunsaturated lipids for low-temperature survival. Proc. Natl. Acad. USA 90: 6208–6012.

Mizuguchi G, Shen X, Landry J, Wu WH, Sen S, Wu C (2004) ATP-driven exchange of histone H2AZ variant catalyzed by SWR1 chromatin remodeling complex. Science 303: 343–348.

Murakami Y, Tsuyama M, Kobayashi Y, Kodama H, Iba K (2000) Trienoic fatty acids and plant tolerance of high temperature. Science 287: 476–479.

Nakamichi N, Kusano M, Fukushima A, Kita M, Ito S, Yamashino T, Saito K, Sakakibara H, Mizuno T (2009) Transcript profiling of an *Arabidopsis* PSEUDO RESPONSE REGULATOR arrhythmic triple mutant reveals a role for the circadian clock in cold stress response. Plant Cell Physiol. 50: 447–462.

Orvar BL, Sangwan V, Omann F, Dhindsa RS (2000) Early steps in cold sensing by plant cells: the role of actin cytoskeleton and membrane fluidity. Plant J. 23: 785–794.

Peier AM, Moqrich A, Hergarden AC, Reeve AJ, Andersson DA, Story GM, Earley TJ, Dragoni I, McIntyre P, Bevan S, Patapoutian A (2002) A TRP channel that senses cold stimuli and menthol. Cell 108: 705–715.

Penfield S (2008) Temperature perception and signal transduction in plants. New Phytol. 179: 615–628.

Penfield S, Josse EM, Kannangara R, Gilday AD, Halliday KJ, Graham IA (2005) Cold and light control seed germination through the bHLH transcription factor SPATULA Curr. Biol. 15: 1998–2006.

Penfield SD, Kendall S, Macgregor D (2012) Temperature perception and signal transduction: Mechanisms across multiple organisms. In: Tanino KK, Storey KB (eds). Temperature adaptation in a changing climate, CABI Publishing, Wallingford.

Plieth C, Hansen UP, Knight H, Knight MR (1999) Temperature sensing by plants: the primary characteristics of signal perception and calcium response. Plant J. 18: 491–497.

Salomé PA, Weigel D, McClung CR (2010) The role of the *Arabidopsis* morning loop components CCA1, LHY, PRR7, and PRR9 in temperature compensation. Plant Cell 22: 3650–3661.

Sangwan V, Foulds I, Singh J, Dhindsa RS (2001) Cold-activation of Brassica napus BN115 promoter is mediated by structural changes in membranes and cytoskeleton and requires Ca^{2+} influx. Plant J. 27: 1–12.

Sidaway-Lee K, Josse EM, Brown A, Gan Y, Halliday KJ, Graham IA, Penfield S (2010) SPATULA links daytime temperature and plant growth rate. Curr. Biol. 20: 1493–1497.

Sheldon, CC, Burn, JE, Perez, PP, Metzger, J, Edwards, JA., Peacock, WJ, Dennis, ES (1999) The FLF MADS box gene: a repressor of flowering in *Arabidopsis* regulated by vernalization and methylation. Plant Cell 11:445–458.

Smith AP, Jain A., Deal RB, Nagarajan VK, Poling MD, Raghothama KG, Meagher RB (2010) Histone H2A.Z regulates the expression of several classes of phosphate starvation response genes but not as a transcriptional activator. Plant Physiol. 152: 217–225.

Sung S, Amasino RM (2004) Vernalization in *Arabidopsis* thaliana is mediated by the PHD finger protein VIN3. Nature 427: 159–164.

Sung S, He Y, Eshoo TW, Tamada Y, Johnson L, Nakahigashi K, Goto K, Jacobsen SE, Amasino RM (2006) Epigenetic maintenance of the vernalized state in *Arabidopsis* thaliana requires LIKE HETEROCHROMATIN PROTEIN 1. Nat Genet. 38: 706–710.

Swiezewski S, Liu F, Magusin A, Dean C (2009) Cold-induced silencing by long antisense transcripts of an *Arabidopsis* Polycomb target. Nature 462: 799–802.

Tähtiharju S, Sangwan V, Monroy AF, Dhindsa RS, Borg M (1997) The induction of kin genes in cold-acclimating *Arabidopsis* thaliana. Evidence of a role for calcium. Planta 203: 442–447.

Townley HE, Knight MR (2002) Calmodulin as a potential negative regulator of *Arabidopsis* COR gene expression. Plant Physiol. 128: 1169–1172.

Trevaskis B, Hemming MN, Dennis ES, Peacock WJ (2007) The molecular basis of vernalization-induced flowering in cereals. Trends Plant Sci. 12: 352–357.

Van't Hoff JH (1896) Studies in chemical dynamics. F. Muller & Co, Amsterdam, 286 pp.

Whalley HJ, Sargeant AW, Steele JF, Lacoere T, Lamb R, Saunders NJ, Knight H, Knight MR (2011) Transcriptomic analysis reveals calcium regulation of specific promoter motifs in *Arabidopsis*. Plant Cell 23: 4079–4095.

Zarka DG, Vogel JT, Cook D, Thomashow MF (2003) Cold induction of *Arabidopsis* CBF genes involves multiple ICE (inducer of CBF expression) promoter elements and a cold-regulatory circuit that is desensitized by low temperature. Plant Physiol. 133: 910–918.

Zhao R, Sun HL, Mei C, Wang XJ, Yan L, Liu R, Zhang XF, Wang XF, Zhang DP (2011) The *Arabidopsis* Ca(2+) -dependent protein kinase CPK12 negatively regulates abscisic acid signaling in seed germination and post-germination growth. New Phytol. 192: 61–73.

Zhu SY, Yu XC, Wang XJ, Zhao R, Li Y, Fan RC, Shang Y, Du SY, Wang XF, Wu FQ, Xu YH, Zhang XY, Zhang DP (2007) Two calcium-dependent protein kinases, CPK4 and CPK11, regulate abscisic acid signal transduction in *Arabidopsis*. Plant Cell 19: 3019–3036.

2 Plant acclimation and adaptation to cold environments

Bob Baxter

2.1 Introduction

Due to their sessile nature, plants, unlike animals, are unable to escape from the many abiotic and biotic factors that cause a departure from optimal conditions of growth and development. As considered elsewhere in this volume, environmental variables, especially those affecting temperature and water availability, are major determinants of plant growth and development. Plants growing at the middle to high latitudes will regularly experience low temperatures, the degree and extent of which will depend upon their location on the Earth's surface. Only *ca.* one-third of the total land area of the Earth is free of ice, and *ca.* 42% of this land experiences temperatures below −20°C (Juntilla and Robberecht 1999; Janska et al. 2010). Thus, plants growing in temperate to high latitudes, and at high altitude at any latitude (the alpine zone), may experience low temperatures on a regular basis throughout their developmental lifespan. Such natural environments contain a wide variety of stress or disturbance factors, for example, cold temperature, limited nutrient supply, occasional drought, intense ice-crystal abrasion, and destructive soil movement processes such as frost heaving. In such instances, the severity of the physical environment produces the primary limitation on plant growth (Callaghan and Jonasson 1995).

Plants growing in cold environments are evolutionarily adapted to cold and freezing temperatures and often (notably, e.g., in alpine environments) high light intensities (Lütz 2010). The nature of such adaptations remains a highly active field of contemporary research, incorporating studies of morphological, anatomical, ecophysiological, cellular, metabolic, and molecular levels of adaptation. In recent years, advances in cellular and metabolic techniques have added greatly to our level of understanding of perception of cold by plants, transduction of cold signals, and changes at the level of the genome that lead to the ecophysiological responses that have long been studied in cold environments. There has never been a better time to integrate such ecological, physiological, cellular, and molecular knowledge to understand adaptation to cold, not

Temperature and Plant Development, First Edition. Edited by Keara A. Franklin and Philip A. Wigge.

least in the wealth of information that can be gained from the study of native plant species to be applied (a) in the further development of crop plants utilized by man and (b) in predicting the likely impacts of global climatic change upon the winter biology of temperate, alpine, and arctic environments.

In the context of agricultural food production, freezing damage is a very serious economic problem, such as in the wheat-growing areas of the Canadian prairies and the Russian steppes, where spring temperatures can be very low. For example, in the province of Saskatchewan, Canada, frost damage to the 2003–2004 crop was estimated at half a billion dollars. Many other such loses are reported regularly across the midlatitude continental agricultural regions of the world. In response, research effort is being directed toward mitigation of such losses, through a better understanding of susceptibility and resistance to cold.

This chapter broadly addresses the key issues related to such susceptibility and resistance in agronomic species as well as wild species that are native to temperate, high-latitude, and high-altitude environments. In addition, it considers the future benefits that such an integrated understanding might afford at a time of rapid change in prevailing winter climate over the coming decades.

2.2 Chilling and freezing injury

Low temperatures may be the direct cause of injuries to plant cells. The so-called chilling-sensitive plants (species native to regions experiencing benign climates, normally with little amplitude in temperature) suffer lethal injuries at temperatures a few degrees above freezing point (typically between *ca.* 10°C and 0°C, but sometimes at higher temperatures (Larcher 2003)). Symptoms of chilling injury are seen in a lowered growth rate, chlorosis of the leaves, loss of turgor (wilting), and, if meristems are compromised, death (Levitt 1980; Srivastava 2002). The degree of cold tolerance can be increased in many cases through a process of gradual acclimation to low, but not freezing, temperatures.

Plants native to cold environments, while being insensitive to cold temperatures above the freezing point, may be damaged at temperatures below that point. Such freezing-sensitive tissues may be killed as soon as ice forms within them. Both chilling and freezing injuries involve damage to cell membranes. However, freezing injury involves additional, more severe, changes involved with the formation of ice. Such ice formation *per se* may not be very damaging, but the resulting dehydration is. Freezing-induced cellular dehydration is, therefore, the most widespread cause of tissue damage in plants. This ice formation usually begins within the apoplast since the solute content here is much lower (and the freezing point therefore is higher) than that of the intracellular fluid (Srivastava 2002). As temperatures fall below 0°C, intracellular ice-crystal formation is initiated in subepidermal and perivascular tissue (Pearce and Ashworth 1992). A degree of extracellular crystallization may be tolerated. However, formation of ice within the apoplast leads to a lowering of water potential, and as a consequence, water moves to the apoplast from the symplast, lowering the water potential of the extracellular fluid, drawing over 90% of the osmotically active water from within the unfrozen cells into the extracellular fluid (Thomashow 1999). This can lead to dehydration damage to both membranes and macromolecules. The membrane damage results from a change in phase from semicrystalline to gel, with an associated loss of activity of intrinsic membrane proteins and enzymes (e.g., ATPases, carrier and channel proteins). In cereals the plasma membrane is the membrane most vulnerable to this type of damage; other membranes are also affected but often at a lower temperature. In other species, damage to the tonoplast limits survival (see the review of Pearce (2001)). The cell contents freeze, resulting in physical shear of the plasma membrane and organelles. This can be particularly prevalent if the tissues experience repeated freeze–thaw cycles, common in autumn–winter and winter–spring transition periods. Such damage is manifest even before thawing of the tissue as a loss of the selective-permeability properties of the membranes and subsequent leakage of electrolytes, phenolics, and a range of other stored solutes (Srivastava 2002).

Given the preceding potential impacts of cold and freezing, it is imperative that plants exposed to low and freezing temperatures either avoid or are able to tolerate the low temperatures that threaten their survival. In the following paragraphs, key plant survival strategies are considered.

2.3 Freezing avoidance and tolerance at the structural and physiological level

2.3.1 Freezing avoidance

Native wild species adapted by natural selection to cold environments have evolved a number of physiological and morphological mechanisms to improve survival in the face of extended cold periods and therefore potential chilling or freezing injury. Such species often exhibit a short stature (graminoids, herbs, and dwarf shrubs), taking advantage of growth in the sheltered, microclimatic boundary layer next to the Earth's surface. This is typified by arctic tundra and alpine species and includes those growing in cushion and rosette forms and in sheltered crevices of alpine and arctic

Table 2.1 Categories of mechanisms associated with the adaptation of plants to freezing temperatures.

Avoidance	Tolerance
Ice nucleators	Compositional changes in membranes
Antifreeze factors	Osmotic adjustments
Preferential sites of ice accumulation	Regulation of plant hormones Cryoprotective compounds
Supercooling	Antioxidant defense systems
Deep supercooling	Cold-inducible proteins
Glass formation	Production of compatible solutes and sugars
Ice barriers	Regulation of acclimation and deacclimation

Modified from Gusta and Wisniewski (2013) © *Physiologia Plantarum* 2012.

landscapes. Many of the responses associated with the aforementioned can be ascribed to either cold avoidance or its tolerance.

Avoidance and tolerance are two key strategies used by plants to deal with a number of abiotic stresses, including those associated with cold and freezing temperatures (Table 2.1). The mechanisms used to survive at such temperatures vary greatly, depending upon the plant species and, in some instances, the tissue or organ that is exposed to cold or freezing (Sakai and Larcher 1987; Gusta and Wisniewski 2013).

Avoidance of cold and freezing temperatures includes life cycle and life-form adaptations, such as those of the cryptophytes (*sensu* Raunkiaer et al. (1934)) that have either resting buds lying beneath the surface of the ground as a rhizome, bulb, or corm or a resting bud submerged under water, thereby avoiding tissue and organ exposure to low winter temperatures; plant growth resumes in the spring from these resting perennating organs. Other growth forms, such as the hemicryptophytes (buds at or near the soil surface) and the chamaephytes (buds on persistent shoots near the ground – e.g., woody plants with perennating buds borne close to the ground, no more than 25 cm above the soil surface), are all dwarf in stature, often protected by winter snow cover (Larcher 2003). The development of prostrate or rosette growth morphology is assumed to be a morphological consequence of development at low temperature (Roberts 1984; see later) and indeed has been used as a selection criterion for cold hardiness in agricultural species over past decades.

Freezing avoidance mechanisms, beyond those of the life-form, are all typically associated with some form of physical attribute of the plant determining whether ice forms, and where it forms, within the plant. The following paragraphs are intended to introduce the reader to the range of such attributes.

2.3.2 Freezing point depression, supercooling, deep supercooling, and extracellular and extraorgan freezing

Many plants found in cold environments accumulate compatible solutes (solutes which can be tolerated in high concentrations) in their intracellular, and often extracellular, solutions. High concentrations of such molecules increase the viscosity and thus reduce diffusion in solutions. This slows metabolism but has advantages for slowing cellular dehydration and for vitrification (see Section 2.3.4) (Wolfe and Bryant 1999). High cellular solute concentrations depress the freezing point of the cell contents.

Supercooling refers to taking a liquid below its equilibrium freezing point, without freezing. This freezing avoidance mechanism has the advantage that the solutions remain liquid and allow relatively normal, though slower, metabolism. Plant leaves often supercool a few degrees and may thus survive mild frosts without freezing damage, whereas slightly colder temperatures can cause extensive damage (Lutze et al. 1998). Freezing avoidance by deep supercooling of tissue water to near its homogeneous nucleation temperature (approximately −40°C) has been shown to be an important survival mechanism in reproductive and vegetative parts of many winter-hardy plants (Pearce 2001). Deep supercooling is usually employed by limited tissues (xylem ray parenchyma, buds, seeds) within a plant and other tissues; leaves and living bark, for example, undergo extracellular freezing upon exposure to prolonged subzero temperatures (Ishikawa 1984). Xylem and buds in extremely hardy twigs which tolerate −70°C are considered to owe their cold hardiness mechanism to extracellular freezing and extraorgan freezing (translocated ice formation), that is, the presence of ice exclusively in the regions of the tissue outside the cell. Here there is movement of water from the tissues to extracellular spaces, where it freezes in extensive masses. The cell cytoplasm has a higher solute concentration, and as a result, intracellular freezing is delayed.

Even given all the advantages of supercooling outlined earlier, such supercooled solutions are intrinsically unstable in nature, and ice formation results if an ice crystal is introduced to the solution; this is always a distinct possibility in biological solutions, via the process of ice nucleation.

2.3.3 Ice nucleation and structural and thermal ice barriers

Plants and plant parts freeze when they cannot avoid nucleation and cannot prevent the growth of ice. In nucleation, water molecules come together to form a stable ice nucleus, either spontaneously (homogeneous nucleation) or when catalyzed by another substance (heterogeneous nucleation). Such substances include ice nucleation-active bacteria, other biological molecules and structures, plus organic and inorganic debris. Nucleators may be on the plant surface (extrinsic) or within the plant (intrinsic). Extrinsic nucleators (e.g., bacteria on leaf surfaces, such as strains of *Pseudomonas* and *Erwinia*) produce proteins that, under the right conditions, instigate nucleation. Such extrinsic nucleators can only effect nucleation if there is moisture present; dry surfaces will not lead to nucleation. If cooling is rapid, ice may form within the plant cells. Crystallization of the water inside the cell may occur (intrinsic nucleation). Such intrinsic ice nucleation for most plant tissues begins internally on the surface of cell walls, in water transporting elements, or on external surfaces (Guy 1990). In a study of a wide range of alpine species, Hacker and Neuner (2008) reported that in woody dwarf shrubs, herbs, rosette, and cushion plants, a single ice nucleation anywhere in the plant was usually sufficient to result in freezing of the leaves and stems of the whole individual. This suggests that for a range of species, no anatomical barriers to ice propagation exist in the vegetative plant. In the case of alpine graminoids (Hacker and Neuner 2008) and again in lowland cereal graminoids, ice nucleation is required in each separate leaf for ice propagation since there is no connection of the vascular tissues between the various leaves comprising the tussock. These structural ice barriers prevent rapid propagation from one leaf throughout the whole tussock. In addition, ice barriers often slow or prevent ice growing from, for example, stems into the reproductive tissues such as flowers or fruits (see, e.g., Wisniewski et al. 1997). Indeed, this has been shown to be the case for both alpine and lowland spring-flowering plant species (see, e.g., Neuner and Hacker (2012)). In such species reproducing in the spring with likely exposure to freezing temperatures, such an ice barrier adaptation is crucial for survival of the reproductive organs. The structural nature of these ice barriers remains under investigation.

In addition to structural ice barriers, thermal ice barriers also exist in certain plant morphologies. For example, in alpine cushion plants, a thermal gradient is seen to build up during freezing (Hacker et al. 2011) with the flower heads being coldest and the vegetative shoots within the compact cushion form of the plant being significantly warmer. Even under pronounced radiative cooling conditions, these shoots may be warm enough to prevent freezing. When ice nucleation occurs in a single flower, it is restricted to that flower by the thermal barrier, preventing propagation to others via the vegetative shoots. Ice nucleation events are thus required in each separate flower, and so the chance of survival of an individual flower is consequently increased.

In terms of plant development, it is perhaps the lack of sufficient ice barriers against extrinsic ice nucleation that partly explains the susceptibility of early developmental stages (germinating seeds and seedlings) of alpine and arctic plant species, despite older developmental stages having this ability. Recent evidence suggests that this may perhaps be explained by the fact that, unlike in the epidermis of developing cotyledons and shoots, the rhizodermis (root epidermis) is not able to prevent ice entry and that the root may be in direct content with ice masses within the frozen soil (Neuner and Hacker 2012; Neuner et al. 2013). The interplay between extrinsic and intrinsic potential nucleation and their propagation or moderation continues to receive attention, and for a fuller discussion of ice nucleation, its causes, and consequences, I refer the reader in the first instance to Pearce (2001).

2.3.4 Glass transition (vitrification)

Cells with highly viscous contents, such as any cells dehydrated by growth of extracellular ice, may form a glass (vitrify) rather than freeze (Wolfe and Bryant 1999). Such vitrification has been demonstrated to occur especially in a range of boreal species, including poplar (Hirsh et al. 1985) as well as the evergreen boreal conifers that have been shown to survive immersion in liquid nitrogen at −196°C, provided they are first slowly cooled to −20 to −30°C. The cell contents go through a *glass transition* (or vitrification) in this temperature range, whereby the water and dissolved substances in the cell become locked in a molecular suspended animation that prevents cell damage at lower temperatures.

Vitrification can occur in biological systems at ambient temperatures (desiccation) or subzero temperatures (cooling) and has been suggested as a mechanism for membrane protection during dehydration in both these cases. The presence of a glass is suggested to elicit three things in protecting cell membranes: (a) limit further dehydration when formed, (b) lower the probability of ice-crystal formation, and (c) may allow the membranes to remain in the fluid lamellar phase at hydrations and temperatures that normally would lead to deleterious phase transitions (Wolfe and Bryant 1999).This latter point is of great importance, as changes in membrane fluidity will affect many aspects of metabolism at low temperature including photosynthesis, respiration, transport across membranes, vesicle dynamics, organelle biogenesis, and cell division and expansion (Stitt and Hurry 2002).

2.3.5 Antifreeze factors

Plants produce factors that can inhibit ice formation or its growth. Such factors include the so-called antifreeze proteins. These proteins (a low-temperature-responsive gene family), often present in plants in relatively high concentrations, exert an ability to inhibit ice-crystal growth by preventing the accretion of water molecules to the growing ice crystal. By adsorbing onto the surface of the ice crystals, such proteins modify ice-crystal shape and growth (smaller and slower-growing crystals, leading to reduced tissue damage). Recent evidence also reveals that in certain cases such as in overwintering monocotyledonous species, there is a marked similarity of such antifreeze proteins to the pathogenesis-related proteins responsible for resistance against pathogen attack. The latter have also been found to exhibit antifreeze activities (Seo et al. 2008). While being unlikely to prevent freezing *per se* in plants, evidence continues to mount that both these protein types have the dual function of protection against damaging ice formation speed and extent, including controlling sites of ice formation or inhibiting ice recrystallization, as well as protecting against pathogen attack during the winter period (Heidarvand and Amiri 2010).

2.4 Freezing tolerance

Unlike the essentially physically based attributes of the plant in freezing avoidance, tolerance mechanisms are predominantly biochemical adaptations arising from, or regulated by, a specific set of genes (Gusta and Wisniewski 2013). Thus, freezing tolerance exhibits a complex genetic basis, relying upon a coordinated set of physiological and biochemical modifications undertaken during the process of cold acclimation (Heidarvand and Amiri 2010). Biochemical changes induced by cold include the accumulation of cryoprotective sugars, free amino acids, and the expression of cold-regulated (COR) genes (see Sections 2.4.2–2.4.5). Further, protection of cell membranes against freezing–dehydration damage is a major factor in freezing tolerance. This is likely achieved both by changes in membrane lipid composition and by accumulation of substances in the cytosol (see Section 2.3).

2.4.1 Cold acclimation (hardening)

Cold acclimation is the process by which certain plants, upon exposure to low but nonlethal temperatures (usually above 0°C), increase their capacity to survive at low temperatures and gradually become increasingly tolerant to subzero temperatures. The primary function of cold acclimation is to stabilize the cell membranes against freeze injury. The yearly cycle of lowering of temperatures above freezing in late autumn and early winter is the trigger for such acclimation in the natural environment. It is a complex, many-faceted process involving temperature and potentially both light quality and photoperiod and involves changes in the increased expression of many genes, metabolism, and morphology (Browse and Xin 2001).

Cold acclimation is the key to freezing tolerance, but only those plants that are genetically competent to do so acclimate to cold (Srivastava 2002).

The major scientific challenge remains the disentangling of those processes that are critical in engendering freezing tolerance from the many processes that merely respond to low temperature (Browse and Xin 2001; Stitt and Hurry 2002). Previous research has demonstrated the difficulty in determining which processes are affected most severely by cold, with differential responses generating complex, indirect effects.

2.4.2 Genes and regulatory mechanisms in cold acclimation

Plant sensing of temperature is dealt with elsewhere in Chapter 1 of this volume and will not be discussed in any further detail here. Nevertheless, the downstream processes resulting from such temperature sensing are of vital importance in understanding, for example, plant developmental responses to cold. Stress-induced genes can be divided into two major classes: genes involved in stress tolerance and genes required for signal transduction. Stress-tolerance genes enable plants to cope with the stress situation, in terms of both short- and long-term responses; these include genes controlling the synthesis of chaperones and osmolytes and the production of other protective compounds (Tuteja et al. 2011). The advent of large-scale transcriptome profiling has proved extremely valuable in identifying several signal transduction pathways leading from stimulus to an end response. Such studies have revealed a genetic regulatory network underlying plant response to adverse conditions such as cold (Fowler and Thomashow 2002). While there are some signaling pathways that appear stress specific in nature, there are others that exhibit extensive crossover or cross talk which may overlap between pathways (Chinnusamy et al. 2004). This profiling approach has led to the identification of hundreds of genes that encode transcription factors that are differentially expressed under various environmental stresses, such as cold (Shameer et al. 2009).

Much of our current understanding of the regulation of the cold-acclimation response stems from work in the model plant system of *Arabidopsis*. Key players in this molecular response have been found to be the C-repeat binding factor/dehydration-responsive element-binding factor 1 (CBF/DREB1) transcription factors (Zhou et al. 2011). Such transcription factors are induced in response to low, nonfreezing temperatures and, in turn, activate gene expression (COR genes) whose products effect the changes needed for cold acclimation (Thomashow 1999). Recent research has shown that certain of these transcription factors may serve as major 'regulatory hubs' controlling the expression of a large number of genes (Thomashow 2010).

In most systems studied to date, CBFs occur as multigene families. In *Arabidopsis* the family of CBFs which regulates cold gene expression comprises three members, CBF1, CBF2, and CBF3 (Gilmour et al. 1998). The importance of such CBF/DREB1 transcription factors in terms of freezing tolerance in many species has been demonstrated by overexpressing them in *Arabidopsis*, leading to the constitutive expression of *COR* genes and constitutive freezing tolerance, in the complete absence of physiological cold acclimation. These results suggest that transcriptional regulation is an important protective mechanism against cold and other stresses such as drought. It has been shown, for example, that expression of the CBF regulon at noninductive temperatures by the ectopic overexpression of CBF3 leads to extensive changes in the metabolome that largely mirror those that occur in response to low temperature in wild-type (WT) plants (Cook et al. 2004; Guy et al. 2008).

While overexpression of *CBF* family members leads to an enhanced freezing tolerance, *Arabidopsis* transformants have been shown to grow poorly at normal growth temperatures (Kasuga et al. 1999). Similarly, when CBF genes from birch (*Betula pendula*) were expressed in *Arabidopsis thaliana*, these plants exhibited CBF overexpression morphologies, that is, stunted growth (Welling and Palva 2008). Furthermore, research testing the function of CBFs from highly freezing-tolerant members of the genus *Vaccinium* growing in the subarctic (Oakenfull et al. 2013) has also demonstrated this phenomenon. Transgenic *Arabidopsis* overexpressing *Vaccinium myrtillus* (bilberry) CBF was more freezing tolerant than either WT or transgenic *Arabidopsis* expressing CBF from two further *Vaccinium* species (*Vaccinium vitis-idaea* (cowberry) and *Vaccinium uliginosum* (bog whortleberry)), with the growth phenotype associated with CBF overexpression, dwarfism, also being evident (Figure 2.1). This developmental altered (reduced) growth has been demonstrated, at least partially, to be mediated by the DELLA growth regulation system. DELLAs are gibberellic acid-opposable endogenous plant growth inhibitors that act (at least in part) by interfering with the activity of growth-promoting transcription factors (Harberd et al. 2009). The latter author suggests such growth inhibition might have adaptive significance when environmental impacts threaten resource limitation. Prioritization of resource allocation may result in resources being diverted away from growth in the adoption of a strategy that involves reduced resource consumption during a period of wait for improved environmental conditions. Acclimation to cold might be one of a number of such adaptations. Much work remains to be done in this exciting and rapidly growing area of plant development.

Current understanding of the complex nature of acclimation at the molecular level is also being rapidly advanced by the discovery and deployment of a number of important mutants of plants such as *Arabidopsis*. For example, vast screening programs, such as those of McKown et al. (1996), Warren et al. (1996), and Xin and Browse (1998), have identified genes that play important roles in acclimation to cold, rendering mutants that are more freezing tolerant than WT plants without cold acclimation. One such gene is *eskimo1* (*esk1*) which increases freezing tolerance of both non-acclimated

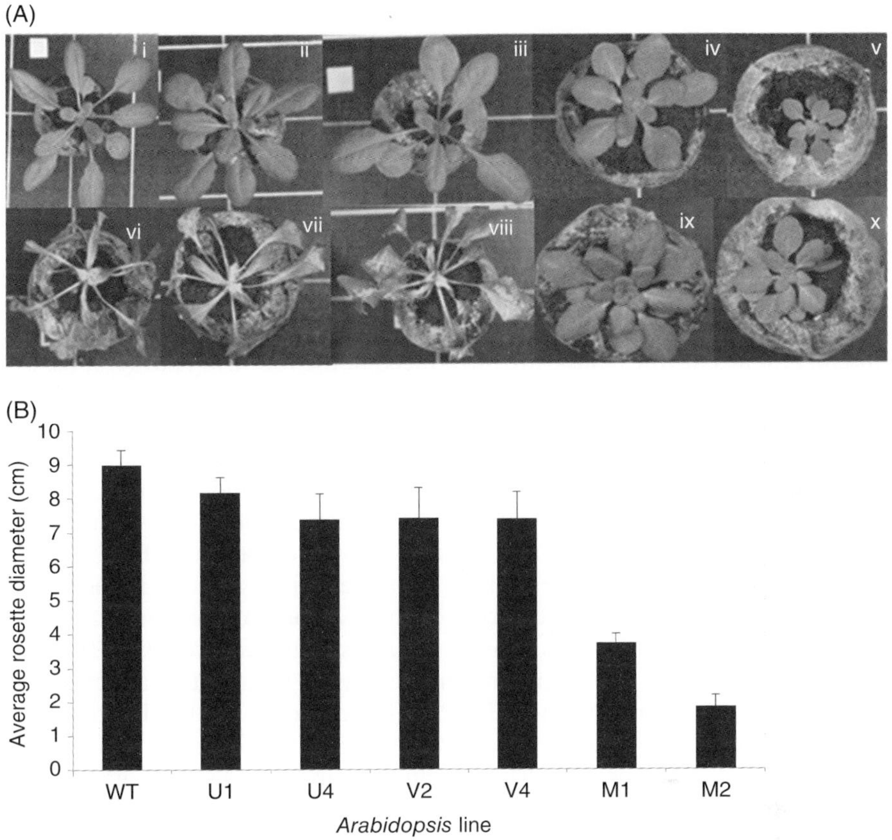

Figure 2.1 Effect of *Vaccinium* CBF overexpression upon freezing tolerance and development in transgenic *Arabidopsis*. (A) Photographs show transgenic *Arabidopsis* expressing *Vaccinium* CBF: *V. myrtillus* (iv, v, ix, x), *V. uliginosum* (ii, vii), *V. vitis-idaea* (iii, viii), and WT (i, vi), before (i–v) and after (vi–x) freezing (−7°C for 24 h in the dark). (B) Bar chart showing average rosette diameter of 'WT' *Arabidopsis* and lines overexpressing either *V. myrtillus*, *V. uliginosum*, or *V. vitis-idaea* CBF ('M', 'U', and 'V', respectively), $n = 8$, error bars are standard errors of the mean. Reproduced from Oakenfull et al. (2013) and Creative Commons. For color detail, please see color plate section.

and cold-acclimated plants when compared to the WT. Other examples are the *SFR* (sensitivity to freezing) genes that confer lower freezing tolerance relative to WT in *SFR* mutants. The different mutation phenotypes exhibit different responses and to differing extents. Work in recent years has progressed in pursuance of determination of the nature and functions of such genes to provide a better holistic understanding of acclimation to cold.

The cold-responsive genes discussed in the preceding text and others that are under current intensive research (see, e.g., Thomashow 2010; Knight and

Knight 2012 and references therein) serve to encode a diverse set of proteins, including enzymes that are involved in the respiration and metabolism of carbohydrates, lipids, antioxidants, antifreeze proteins, molecular chaperones, and others that have a presumed function in tolerance to freezing through metabolic and structural adjustment (Heidarvand and Amiri 2010), for example, active readjustment of the metabolome (Guy et al. 2008). Changes in proteins arise from those metabolites that are differentially expressed in response to cold stress. They are involved in a range of processes, including photosynthesis; photorespiration; metabolism of carbon, nitrogen, and sulfur; and redox homeostasis (Yan et al. 2006). Some of these cold-associated and inducible proteins are the dehydrins, heat shock proteins, antifreeze proteins, and metabolic enzymes; these are introduced in the following paragraphs.

2.4.3 Dehydrins

Dehydrin proteins (group 2 late embryogenesis abundant proteins) are produced as a plant response to those environmental conditions with a dehydrative component, including that induced by low temperature. For example, in barley some 13 dehydrin genes have been identified, with the induction of two (Dhn5 and Dhn8) being detected at the level of transcription. In winter and intermediate, compared to spring, cultivars, a higher level of DHN5 was detected in tandem with a higher freezing tolerance; spring cultivars exhibited a lower level of accumulation of DHN5 (Kosova et al. 2008). Despite such studies, at present the exact function of dehydrins remains unclear. Evidence for their role as stabilizers of cell membranes, in protection of proteins against dehydration, and as *in vivo* antifreeze effectors, as well as potential osmoregulators and radical scavengers, continues to be reported (Hanin et al. 2011).

2.4.4 Heat shock proteins

Heat shock proteins, despite their name, have also been shown to respond to low temperature (particularly HSP90, HSP70, several small HSPs, and chaperonins 60 and 20). Such proteins are molecular chaperones, and it has been suggested that their upregulation may play a vital role in facilitating refolding in denatured proteins and in preventing their aggregation in plants subject to cold (Yan et al. 2006).

2.4.5 Enzymatic and metabolic response in cryoprotection

As outlined earlier, the cold response of plants involves many enzymes and metabolites that are altered by extensive modification of the transcriptome, proteome, and metabolome. Of particular note here is the transcription level of genes that are negatively – *versus* those that are positively – correlated with freezing tolerance. The former includes photosynthesis, tetrapyrrole synthesis, and cell-wall, lipid, and nucleotide metabolism, whereas the latter includes carbohydrate, amino acid, and secondary metabolism (Hannah et al. 2006). Cold stress has been demonstrated to induce the biosynthesis of flavonoids, anthocyanins, and phenylpropanoids, among others (Kaplan et al. 2007). For example, across a wide spectrum of species, anthocyanin and flavonol concentrations in leaves have been seen to rise significantly on acclimation of plants to cold. Together these compounds serve to prevent the overexcitation of chlorophyll under extreme cold conditions (Hannah et al. 2006).

Recent metabolite-profiling studies have served to refocus attention on the aforementioned and other potentially important components that are found in the 'temperature stress metabolome'. One such prominent component of the reprogramming of the metabolome at low temperature is central carbohydrate metabolism, including the determination of a key role for metabolite transporters in carbohydrate metabolism at low temperatures, together with their partitioning between cytosol and the chloroplast (Guy et al. 2008; Usadel et al. 2008). For example, starch breakdown has been demonstrated to play a key role in the autumn–winter transition and the development of low temperature tolerance by supporting carbohydrate metabolism for the production of cryoprotectants and energy generation following the cessation of photosynthesis. However, following an acclimation period of sufficient length, energy constraints may no longer limit the acquisition of freezing tolerance, suggesting that other, genetically determined limitations may operate. Of note here is the mounting evidence that the low-temperature response clearly exhibits a nonlinear behavior over time at the metabolite level (Guy et al. 2008).

Nitrogenous compound metabolism also responds to low-temperature stress (Usadel et al. 2008), notably of certain amino acids and polyamine compounds, including especially those involved in proline biosynthesis (Davey et al. 2009), a key compatible solute in response to low temperature, drought, and salinity.

2.4.6 The role of hormones in low-temperature acclimation

Although the sensing and signaling of temperature are dealt with in detail elsewhere in this volume, it is worth reemphasizing that, as endogenous factors, hormones play a vital role in regulating plant developmental processes. Growth plasticity plays a major role in adaptation of plants to environmental changes (Rahman 2013). It is a highly complex process combining interaction of different hormones at transcriptional, translational, and cellular levels (Chandler 2009; Rahman 2013). Early research on the environmental cues that induce cold acclimation led to the suggestion that abscisic acid (ABA) is involved (see Gusta et al. (2005) for a full description). ABA activates a wide range of genes associated with low temperatures, drought, salinity, desiccation, dormancy, germination, arrest of embryonic development, and the closure of stomata. Evidence to date suggests that ABA can substitute for the environmental cue of a low-temperature stimulus, provided that there is an adequate supply of carbohydrate available. Evidence also suggests that there may be ABA-dependent and ABA-independent pathways involved in the cold-acclimation process, with ABA regulating many of the genes associated with an increase in freezing tolerance.

Here I have highlighted some of the key components of acclimation, but space precludes a comprehensive analysis. I refer the reader to the excellent reviews of Browse and Xin (2001), Penfield (2008), Thomashow (2010), and Knight and Knight (2012), addressing detailed aspects of the molecular basis of plant cold acclimation.

2.5 Cold deacclimation (dehardening) and reacclimation (rehardening)

In climates with a strong seasonal amplitude in temperature, plants that have undergone seasonal cold acclimation in the autumn to afford freezing tolerance throughout the winter period undergo deacclimation (i.e., lose their freezing tolerance) in the spring prior to resumption of active growth. Cold acclimation and deacclimation are necessary for successful freezing tolerance. Such tolerance demands both accurate timing and rates of the acclimatory and deacclimatory processes (Suojala and Linden 1997). While the regulation and physiological, biochemical, and molecular aspects of acclimation (discussed in the preceding text) are now relatively well known, these same aspects in the process of deacclimation are much less so, in terms of regulation both by environmental signals and by associated physiological and biochemical changes (Kalberer et al. 2006). They are, however, as we shall see later in this chapter, of some key potential importance in determining plant survival in a warming climate.

Deacclimation refers to a reduction in those levels of cold or freezing tolerance that were originally attained through an earlier acclimation process, as well as to mechanisms that mediate reduced tolerance. The term is used to describe losses of tolerance invoked by environmental stimuli (spring warming) as well as phenological changes and changes brought by the reactivation of active plant growth (Kalberer et al. 2006). Unlike acclimation (of weeks to months in duration and requiring considerable energy; Browse and Lange 2004), deacclimation has been shown under both laboratory and field conditions to be relatively fast (days to weeks in extent) and to require less energy. These key differences in kinetics have been discussed in the literature in terms of the large changes in structure and function needed upon acclimation and the fact that deacclimation involves more downregulation of gene expression and biosynthesis and may be fuelled by metabolite catabolism (stress proteins, compatible solutes, etc. that often are either synthesized or accumulate during cold acclimation).

Woody perennials exhibit two forms of deacclimation, termed 'active' and 'passive'. Active deacclimation occurs as a response to substantial increases in ambient temperature and is associated with wide-ranging structural and functional changes associated with resumption of growth and progresses rapidly upon inception. It is usually associated with large-scale changes in gene expression. Active deacclimation typically occurs in spring but may occur prematurely in winter as a response to transient warm periods. Passive deacclimation, on the other hand, results from the exposure of fully acclimated plants, in midwinter, to small to moderate elevations (5–8°C or less) in temperature for extended durations of time. This type of deacclimation is largely associated with depletion of carbohydrate reserves resulting from enhanced metabolism (Kalberer et al. 2006).

Prevention of premature deacclimation in environments where the winter temperatures and spring transitions from cold to warm are unpredictable is highly desirable. A key question however is how is this achieved? The short answer is that it is a complex situation, involving many potential facets. Across a wide range of species, both native and agricultural, it is known that a high degree of midwinter cold acclimation and a high resistance to deacclimation are two different attributes that are independently inherited and not necessarily both present together in plants that have evolved in cold climates. Nor is the ability to acclimate rapidly always associated with a high cold-acclimation capacity or high deacclimation resistance. Current thinking includes the possibility that the plant's resistance to deacclimation is related to the degree of temperature fluctuation (both magnitude and frequency) to which plants in their native habitat are exposed, rather than low temperatures *per se*. (Kalberer et al. 2006, 2007). The latter authors also caution against thinking that low deacclimation resistance is always deleterious to winter survival. If the plant, upon deacclimation, can quickly reacclimate (recover some or all of the lost cold resistance), then freezing damage may not result. Such reacclimation capacity has been found to be a common process in many overwintering plants. It occurs when deacclimated plants are subsequently exposed to cold temperatures.

Spring development of some low-stature alpine and subalpine plants is highly influenced by the timing of snowmelt. Immediately after melt (or earlier for some species that begin growth and development under the melting snowpack), many plants rapidly deacclimate. Thereafter and during the spring thaw period and subsequently the growing season, there may be repeated frosts, leading to repeated freeze–thaw cycles. It has been demonstrated that temperature fluctuations, such as freeze–thaw cycles, impact upon cold resistance status differently compared to constant temperatures, due to their effect on the deacclimation–reacclimation cycles. The resulting level of cold resistance from such alternating warm and cold cycles often depends upon the magnitude and duration of each temperature. For example, cold night temperatures can promote reacclimation so long as the daytime temperatures are not too high or too long in duration (Leinonen et al. 1997). The capacity for reacclimation is dependent upon the magnitude or duration of warm temperature exposure as deacclimation progresses. The longer and greater the warming is, the lower is the reacclimation potential of the plant.

It is important to acknowledge at this point that avoidance and tolerance mechanisms are not mutually exclusive in their contribution to overall plant survival of cold. They work in tandem, with both being under genetic control and both evolved in response to environmental selection pressures. As Gusta and Wisniewski (2013) clearly state: 'The ability of a plant to segregate ice into specific areas of its tissue where it will do no harm may be as critical as the ability of its cells to withstand the deyhdrative stress associated with the formation and presence of that ice'.

Having discussed the key avoidance and tolerance strategies, I turn now to a consideration of the importance of an understanding of issues of space and time in relation to an understanding of plant response to low temperature.

2.6 Spatial and temporal considerations of plant responses to low temperature

Mechanisms that are responsible for determining the avoidance or tolerance of plants to cold may differ within the same plant over short distances (e.g., root *vs*. shoot tissues or, in woody plants, the xylem tissue *vs*. the bark) and even between adjacent cells (Gusta and Wisniewski 2013). In this whole-plant or whole-tissue context, there are likely different degrees of cold stress exerted in different tissues at differing positions within the plant. Adaptations of these different tissues and cells may thus show differences spatially.

Just as tissue type and its location are important *in vivo*, so too are the specific timing and extent of the cold or freezing event(s) and the ontogenetic stage of the plants themselves. For example, in the natural environment, plants may experience either acute or chronic cold stresses. The former can be characterized by the early or late hard frosts experienced in early to late autumn or spring periods or, at high latitudes and high altitudes, heavy frosts occurring even in summer (Sierra-Almeida and Cavieres 2012), giving rise to acute freezing stress events. Chronic freezing stress, on the other hand, can be for periods of weeks to months throughout the winter season in the higher Northern latitudes and at high altitudes.

The risk of frost damage in winter is, however, actually rather low for prostrate plants that are covered permanently by snow (at constant temperatures between 0°C and −5°C (Larcher et al. 2010), but in exposed, windblown, snow-free localities, the plants must be fully frost resistant to temperatures often at −30°C or lower. These same plants, whether winter snow covered or not, must also be able to withstand very high summer temperatures (30–40°C) resulting from direct solar insolation (i.e., an annual amplitude of some 60–80°C).

Observation of the recent literature reveals that despite many observations and measurements being made across a wide range of species, little is still known about the freezing resistance changes observed and measured over the course of a year in extreme environments (Larcher et al. 2010). Furthermore, when considering plant stature and size, there are further contextual issues that must be borne in mind. In high-elevation habitats, for example, an apparent trade-off between high growth rates and freezing resistance has led some authors to assume that seedlings are less freezing resistant than adults. However, it should be noted that juvenile seedlings of, for example, trees may, by their short stature, be exposed to lower temperature and longer freezing events near the ground if they are not snow covered (the classic ground frost). They could therefore actually be more freezing tolerant than their taller adult counterparts (Sierra-Almeida and Cavieres 2012). A greater understanding of such potential trade-offs related to the acclimation and deacclimation phenomena discussed earlier will further facilitate our overall comprehension of plant response to cold throughout their ontogenetic development in both natural and agronomic systems. In the following paragraphs the current status of knowledge related to the interactions of key environmental stresses concomitant with cold is assessed.

2.6.1 Interactions between cold and light: Winter dormancy

Perennial plants growing in temperate and high-latitude regions anticipate the approach of winter by sensing the associated reduction in day length. When the day length falls below the critical length permitting growth, cell division terminates in the meristems (Nitsch 1957). Continued exposure of these so-called 'ecodormant' plants to short days results in the transition from a state of 'ecodormancy' to one of 'endodormancy' (characterized by an inability of the meristems to respond to growth-promoting signals, in contrast to the ecodormant state) (Baba et al. 2011). To restore the ability of endodormant meristems to respond to growth-promotive signals and to reinitiate growth subsequently requires exposure to chilling temperatures.

2.6.2 Interactions between cold and environmental drought

At low temperatures the increased viscosity of water and decreased hydraulic conductivity of root cell membranes lead to a reduced water uptake. Plants adapted to growth and development at high latitude and altitude can still extract water from soils even near freezing point, whereas species growing at lower latitudes often struggle below 10°C (Larcher 2003). Frost-induced drought (winter desiccation) is a key danger in alpine and high-latitude plants (Körner 2003; Crawford 2008), especially for evergreen species, where bright sunlit days warm the leaves, but the roots and parts of stems remain frozen, with no liquid water availability, leading to partial lethal damage to leaves. This can lead to overheating of the plant crown and loss of water by evaporation. This, in turn, leads to 'frost drought', a major component of which is cavitation of air bubbles within the xylem and formation of an embolism, breaking the water column connectivity that is vital for the maintenance of hydraulic conductivity and the efficacy of cooling related to the transpiration. Despite the widespread importance of such 'frost drought', there remains a dearth of information and relevant datasets by which to study its intraspecific variability (Charra-Vaskou and Mayr 2011).

2.6.3 Interactions between cold and light: Photosynthesis, photoinhibition, and reactive oxygen species in cold environments

As we have already seen, the process of cold acclimation in plants is complex and is not only dependent upon growth temperature and the developmental stage of the plant prior to the freezing event. It is also dependent upon the irradiance at which the plants are grown (Gray et al. 1997), which may be photoinhibitory (the light-dependent reduction in photosynthetic efficiency). Such photoinhibition may be a consequence of either the irreversible light-induced inactivation of photosystem II (PSII) reaction centers or the reversible downregulation of PSII through the non-radiative dissipation of excess light. The combination of low temperature and high solar irradiance has been associated with low-temperature photoinhibition in both agricultural (reviewed in Baker et al. 1994; Krause 1994) and native species (e.g., Ball 1994). The photosynthetic response of plants at low temperature is dependent upon time of exposure and the developmental history of the leaves (Huner et al. 1993). Here it is also pertinent to consider reactive oxygen species (ROS), such as superoxide anion radicals ($O_2^{\cdot -}$), hydrogen peroxide (H_2O_2), and hydroxyl radicals ($OH\cdot$), that are unavoidable by-products of oxygenic photosynthesis, causing progressive oxidative damage and ultimately cell death if not dealt with by the plant. Chloroplast membranes are in particular sensitive to oxidation stress damage caused by the generation of excessive amounts of ROS in these membranes. ROS can cause extensive peroxidation and de-esterification of membrane lipids, as well as lead to protein denaturation and mutation of nucleic acids (Bowler et al. 1992). Environmental stress conditions, including low temperature, exacerbate the production of ROS as a toxic by-product. These ROS, in turn, lead to oxidative stress. Detoxification of harmful ROS is carried out by a highly efficient antioxidant system, consisting of both non-enzymatic (e.g., ascorbic acid, reduced glutathione, flavonoids) and enzymatic antioxidants (e.g., superoxide dismutase, catalase, guaiacol peroxidase) present in plant cells. In the context of winter cold tolerance and the susceptibility of woody plants to extreme frost injury at the end of winter after a warm period and late spring frosts, Pukacki and Kaminska-Rozek (2013) studied the seasonal changes in reactive species, antioxidants, and cold tolerance during deacclimation of populations of Norway spruce (*Picea abies*) taken from locations across its natural geographic range. These authors confirmed earlier findings that cold-acclimated plants repair the damage caused by free radicals much more rapidly than deacclimated plants do and reported that the activity of the antioxidant systems was related to the rate of photosynthesis and to freezing tolerance. Their results also indicated that genetic differences between populations affected the antioxidant system less strongly than did the changing climatic conditions at the site where the trees were cultivated, suggesting that the response of the antioxidant system depends more strongly on climatic conditions than on population origin.

Low temperatures inhibit photosynthesis by increasing the CO_2 diffusion resistances within leaves when solutes freeze (−5 to −9°C) and by directly affecting photosynthetic membrane and biochemical function (Kappen 1993; Larcher 1995). However, some high-latitude vascular plants photosynthesize at temperatures below −4°C, and tundra evergreen species are photosynthetically active immediately upon emergence from the snow (Semikhatova et al. 1992). The subnivean environment is characterized by low light levels and temperatures far below freezing for most of the winter (Woolgrove and Woodin 1996).

Activation of the photosynthetic apparatus under snow also enables the plants to reach their maximum capacity quickly once melt has occurred (Kudo et al. 1999; Starr and Oberbauer 2003). This activity allows evergreens to rapidly increase photosynthesis upon snowmelt and reduces wintertime losses of carbon which may be especially important in, for example, arctic ecosystems (Körner and Diemer 1987). In the case of the latter, those species growing at high altitude in the alpine zone may experience some of the most severe conditions for low-temperature photoinhibition due to the high frequency of frost events and the high sunlight exposures that are characteristic of high elevations (Ball 1994). Plants emerging from snowbanks at these elevations may experience even greater sunlight levels due to the high albedo of snow, as well as lower air temperatures, than other alpine plants (Körner 1999). Germino et al (2002) investigating two such alpine snowbank species (*Caltha leptosepala* and *Erythronium grandiflorum*) in situ reported that despite considerable differences in plant form and environment, *C. leptosepala* and *E. grandiflorum* appeared to have similarly high resistances to low-temperature photoinhibition. The rapid recovery of photosynthetic CO_2 uptake following frost nights and exposure to high photon flux density, plus little response of photosynthesis to warmer nights and shade the following morning, points to a particularly high resistance to low-temperature photoinhibition in these snowbank species. Without such a resistance to cold stress, photoinhibition, and their combination, the brief seasonal opportunity for carbon gain in the alpine, especially in snowbank habitats, would be even less (Körner 1999).

While many physiological studies of plants from alpine environments have been undertaken, there are many fewer studies investigating cellular and tissue structure and function in a plant development context. One such cytological study of over 30 plants growing in cold alpine environments revealed a similarity of most leaf cellular structures when compared to plants growing under more temperate climatic conditions. The only significant structural differences described to date in photosynthetic leaf tissue of plants from cold environments have been in the chloroplasts and, occasionally, in the peroxisomes (Lütz 2010). Such differences in the chloroplast are manifest as 'protrusion' extensions (tubular, extended stroma-filled areas, or increases in envelope surfaces and in stroma volume, allowing a large surface extension of the chloroplast) and appear to develop preferentially in plants from cold environments. They are thought to act as adaptive structures in plants, maximizing photosynthetic productivity in these alpine plant species with a short vegetative period of growth.

From the preceding discussion the reader will appreciate that the correct timing of acclimation, deacclimation, reacclimation, and resumption of active growth in perennials is of vital importance for both their winter and post-winter survival. For further detailed information on acclimation, deacclimation, and reacclimation, I refer the reader to the review of Kalberer et al. (2006) and references therein. With predicted changes in future autumn, winter, and spring climates, a number of impacts upon plant photosynthesis, growth, and development may ensue and it is to those that we now turn.

2.7 The survival of cold and freezing stress in a changing climate

Global climate models predict a number of important climatic changes over the coming decades. These include the increase in mean surface air temperature, resulting from increased atmospheric carbon dioxide concentrations from the burning of fossil fuels and an associated increase in the frequency and severity of erratic temperature events (Bokhorst et al. 2009). Temperate, boreal, and subarctic winters are being demonstrated to be milder (see, e.g., Gu et al. (2008), and Bokhorst et al. (2009)), with rising temperatures, declining snow cover, and increasing precipitation being some of the major trends recorded in the case of the Northern high latitudes. If snow cover depth and duration in alpine, arctic, and temperate environments are reduced through climatic warming, plants that are normally protected by snow-lie in winter may become exposed to greater extremes of temperature and solar radiation (Bannister et al. 2005).

In alpine environments, freezing events rarely kill plants, but they can cause large losses in biomass and significantly reduce reproductive output and in doing so modify the abundance and competitive ability of individual plant species. As we have seen earlier in this chapter, the early growing season is usually the most critical period for freezing damage, where developing new leaf and shoot tissues may be more sensitive than mature tissues (see, e.g., Taschler et al. (2004) and earlier references in the present chapter). Earlier beginning of the growing season due to earlier snowmelt can have multiple consequences. It could increase the length of the photosynthetic period, if the end of the season remains fixed or changes to a later date. However, deleterious effects may also arise if, for example, phenological phases of growth and flower bud are being initiated earlier due to earlier snowmelt, but post-snowmelt frost events remain unaltered (Rixen et al. 2012). Such a paradox of increased frost damage in the face of global warming provides important insights into the adaptive significance of phenology and plant development (see, e.g., Inouye (2008)).

Several studies have demonstrated enhanced freezing sensitivity of plants with an earlier snowmelt and warmer ambient temperatures. Of particular note here are the so-called ‘winter-warming’ events. These unseasonable warm periods, often during late winter, leading to sudden and alternating warming above 0 °C and subzero freezing temperatures, endanger the survival of perennial plants, by removing the insulating snow cover through melt or formation of ice layers from refreezing of partially melted snow (Bokhorst et al. 2009, 2011, 2012). In the latter cases, the authors manipulated late winter snow cover under field conditions using infrared heating lamps over native dwarf shrub plant communities to mimic an extreme winter-warming event of less than 10-day duration and compared this to a natural late winter-warming event across the same landscape. They found strikingly similar impacts of plant freezing damage following both artificial and natural warming events. The warm spells led to fast deacclimation and a rapid decrease in the levels of soluble carbohydrates in plants (Bokhorst et al. 2010). While little is known of the impacts of this on reacclimation ability following such extreme warming events, it is likely that lower carbohydrate reserves may decrease reacclimation ability in these circumstances.

In plant species of wide latitudinal extent, adaptation of plants to their climate of origin may play an important role. For example, in species such as mountain birch (*Betula pubescens* ssp. *czerepanovii*), the rate of deacclimation has been shown to be correlated with accumulation of chilling requirements in the different 'latitudinal' ecotypes. Northern ecotypes, with low chilling requirements, show the highest rates of deacclimation and oceanic or low-altitude ecotypes, with high chilling requirements, were more effective in maintaining cold-acclimation status (Taulavuori et al. 2004). In other species, however, no consistent relationships between the degree of endodormancy and rates of deacclimation have been found (Pagter and Arora 2013).

Elevated atmospheric CO_2 concentrations on their own have also been shown to have the potential to enhance freezing sensitivity of plants across a range of life-forms and developmental stages, including saplings of tree species (Loveys et al. 2006), subarctic shrubs (Beerling et al. 2001), and temperate forbs and grasses (Obrist et al. 2001). For example, Lutze et al. (1998) showed that supercooling of *Eucalyptus pauciflora* leaves could be altered by environmental change: the leaves supercooled to −4.7°C ± 0.5 8°C when the plants were grown in an atmosphere with current CO_2 concentration but only to −3.5°C ± 0.4 8°C when grown in a CO_2-enriched atmosphere. In some instances, however, growth in an elevated concentration of atmospheric CO_2 had either no effect (Taulavuori et al. 2001) or actually reduced freezing sensitivity (Loik et al. 2000). Furthermore, this direct impact of elevated CO_2 may be further complicated by the finding that this may, at least in some instances, be altered by temperature (Woldendorp et al. 2008).

Taken together, the evidence marshalled in the preceding text suggests that multiple factors, including time of season, climatic factors, dormancy status, genotype, and genetic adaptation to the local climate, all impact upon determining the freezing sensitivity of plants. As discussed previously, the ability to reacclimate if low temperatures follow a period of deacclimation may be crucial to plant survival. Recent evidence suggests that some capacity to reacclimate is seen in many plants, this capacity often being greater in early winter and when plants are fully dormant (Pagter and Williams 2011). Further study of plants from environments that regularly face the threat of post-deacclimation frosting events affords an opportunity to incorporate the knowledge gained into breeding of crops for future climates (see Section 2.8).

In summary, a high resistance to deacclimation and a high efficiency of reacclimation may be crucial in plant survival over winter in a changing climate. They remain relatively little studied but are worthy of much greater attention. Pagter and Arora (2013) encapsulate the key points in a very

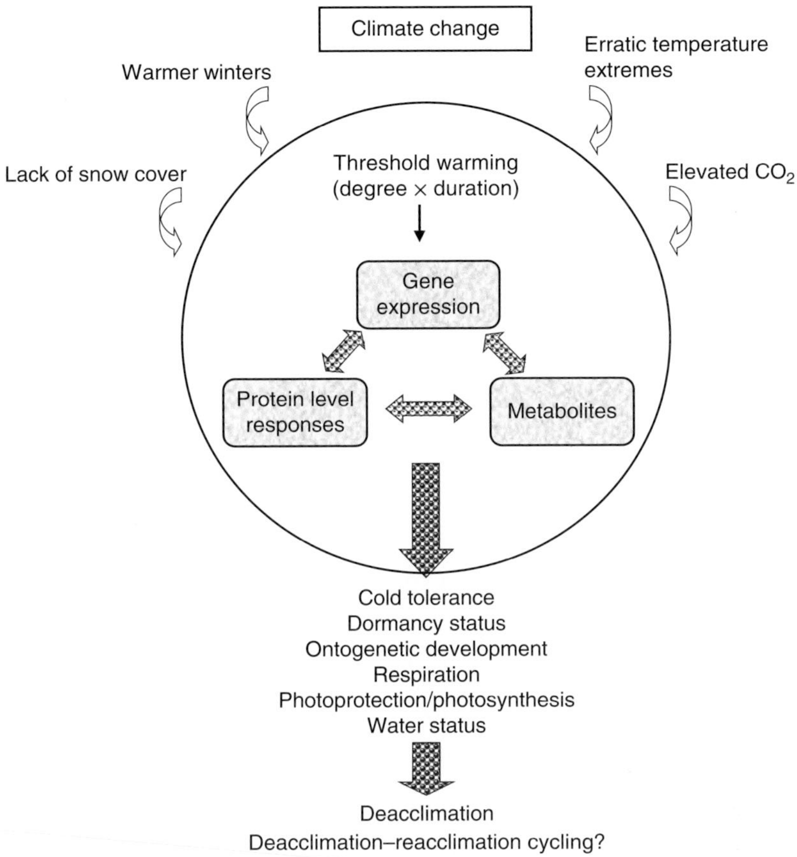

Figure 2.2 Summary model of how winter climate change may affect a range of physiological mechanisms and subsequent deacclimation of perennial plants. Elevated winter temperatures, warm spells, and elevated atmospheric CO_2 concentrations may contribute to the threshold warming (combined effect of both the degree and the duration) required to induce deacclimation. Threshold warming results in altered gene expression, which leads to changes in the proteome and metabolome and associated physiological responses that determine the extent of deacclimation. Some winter climate changes may additionally have a more direct effect on plant physiology and deacclimation; for example, warmer winters can increase respiratory metabolism and cause 'passive' deacclimation. Photoprotection is particularly important in evergreens. Photoprotection (or lack thereof) does not cause deacclimation, but photoprotection and photosynthesis are essential for winter survival and spring growth and are sensitive to lack of snow cover and erratic temperature extremes. If deacclimation is not beyond the reversibility threshold and returning cold is inductive and tissues have reacclimation ability, then reacclimation is possible. Reproduced from Pagter and Arora (2013) © *Physiologia Plantarum* 2012.

useful summary figure (reproduced in Figure 2.2), showing the importance of the postulated interactions between gene expression, changes in the proteome and metabolome, and the associated physiological responses determining the extent of deacclimation in a warmer world.

2.8 Plant cold acclimation and adaptation in an agricultural context

In this century, the world will face the formidable challenge of increasing agricultural productivity without significantly expanding the area under cultivation (Edgerton 2009; Godfray et al. 2010). In mid- and high-latitude regions, cold temperatures impede crops from achieving their productivity potential (Tuteja et al. 2011). It is insufficient cold hardiness that remains a major impediment to reliable crop production in cold climates. Improvement of persistence in harsh winters remains a key objective of breeding programs developing suitable cultivars. However, conventional breeding methods have met with limited success in improving the cold tolerance of important crop plants involving interspecific or intergeneric hybridization. It is becoming very pressing therefore to identify alternative strategies to develop cold-tolerant crops. To date, the fragmentary understanding of the molecular and genetic bases of superior cold adaptation has impeded breeding efforts (Castonguay et al. 2013). As we have seen earlier, plant responses to cold stress are highly complex, so the path ahead is a very difficult one. However, there already exist a range of transgenic lines of different crops which have shown improved tolerance to cold (Sanghera et al. 2011). Although our knowledge of the transcriptional control of the low temperature response is currently limited, it is constantly improving. Nevertheless, a large number of genes identified in different studies are currently annotated with 'unknown function'. The development of genetically engineered plants by the introduction and/or overexpression of selected genes seems to be a viable option to hasten the breeding of plants with improved low-temperature and freezing tolerance. Here, however, we meet not only the problem of acceptability of such approaches by the general public but also the scientific issues of this approach being the only option when genes of interest originate from cross barrier species, from distant relatives, or from non-plant sources (Sanghera et al. 2011).

2.9 Summary

Our knowledge of acclimation and adaptation to cold in plants has grown significantly over the last few decades, notably through the use of cell and tissue cultures and the use of model organisms such as *Arabidopsis* to study genes, proteins, and metabolites in the acquisition of cold and freezing tolerance. Such studies continue to provide much information concerning cold perception, signaling, and responses (Ahuja et al. 2010). This is vital since such low-temperature stress responses are very highly complex multigenic processes. Variation in the level of adaptation to cold can be genetically determined, as well as being affected by plant developmental stage and physiological status at the time of exposure (Janská et al. 2010). In order to fully understand the entirety of the complex interactions, a systems approach, taking full advantage of emerging metabolomic technologies, is highly desirable to allow identification of the full range of metabolites and pathways involved in low-temperature response of plants and induced tolerances. Inextricably linked to this, there remains the fundamental necessity to understand all these findings within the context of the growth habit, the physiology, and the life cycle of whole plants growing under natural environmental conditions (see, e.g., Gusta and Wisniewski 2013). Furthermore, when studying the relevant literature on plant developmental responses to cold, it is readily apparent that there is very often a lack of congruence between studies carried out in controlled, laboratory, environments and those undertaken under field conditions, even when considering the same species.

Major breakthroughs in instrumentation technology are facilitating the measurement and manipulation of plants in situ in their native environment. While much can be achieved in the laboratory, a better understanding of the interplay of the complex of cold and other stresses, to which plants are often subjected *in vivo*, is beginning to pay large dividends. This is of vital importance since in many cold-acclimation studies that have been undertaken under controlled environment conditions, the low-temperature regime continually drives the cold-acclimation reactions. Under natural environmental conditions, temperatures may fluctuate wildly between acclimatory and deacclimatory conditions, perhaps on a daily basis under certain circumstances. The ability to manipulate freeze and thaw in the field is an excellent example of this progressive and interdisciplinary approach. Such studies allow the natural range of adaptation to be incorporated into our understanding of metabolic survival behavior (see, e.g., Lutz 2010). There remains a need for careful integration of approaches at all levels of complexity, from whole plant to the molecular, targeting the key mechanisms of perception, transduction, elicitation, and response to cold, fully acknowledging the importance of environmental space and time and the inevitable heterogeneity that this comprises. In consideration of such future approaches, our studies should seek to encompass the whole range of plant functional types, lest we overlook something of vital importance. For example, in the context of high-latitude and high-altitude environments, despite often comprising a high proportion of the plant life, there remains a dearth of information related to cellular physiological and molecular research on both ferns and mosses growing in cold environments (Lütz 2010).

In my own research in cold environments, I am finding much profitable collaboration between the fields of ecology, physiology, and cell and molecular biology. The multi-way exchange of information, ideas, and experimental protocols in both laboratory and field settings remains a highly exciting prospect, and much is still to be learned for both wild and agronomic plant species from the highly profitable cross-disciplinary understanding of plant adaptation and development in cold climates.

References

Ahuja I, de Vos RC, Bones AM, Hall RD (2010) Plant molecular stress responses face climate change. Trends Plant Sci. 15: 664–674.

Baba K, Karlberg A, Schmidt J, Schrader J, Hvidsten TR, Bako L, Bhalerao RP (2011) Activity–dormancy transition in the cambial meristem involves stage-specific modulation of auxin response in hybrid aspen. Proc. Natl. Acad. Sci. USA 108: 3418–3423.

Baker NR, Farage PK, Stirling C, Long SP (1994) Photoinhibition of crop photosynthesis at low temperatures. In: Baker NR, Bowyer JR (eds). Photoinhibition of photosynthesis from molecular mechanisms to the field. BIOS Scientific Publishers, Oxford, pp. 349–363.

Ball MC (1994) The role of photoinhibition during tree seedling establishment at low temperatures. In: Baker NR, Bowyer JR (eds). Photoinhibition of photosynthesis –From molecular mechanisms to the field. BIOS Scientific Publishers, London, pp. 367–378.

Bannister P, Maegli T, Dickinson KJM, Halloy SRP, Knight A, Lord JM, Mark AF, Spencer KL (2005) Will loss of snow cover during climatic warming expose New Zealand alpine plants to increased frost damage? Oecologia 144: 245–256.

Beerling DJ, Terry AC, Mitchell PL, Callaghan TV, Gwynn-Jones D, Lee JA (2001) Time to chill: Effects of simulated global change on leaf ice nucleation temperatures of subarctic vegetation. Am. J. Bot. 88: 628–633.

Bokhorst SF, Bjerke JW, Tommervik H, Callaghan TV, Phoenix GK (2009) Winter warming events damage sub-arctic vegetation: Consistent evidence from an experimental manipulation and a natural event. J. Ecol. 97: 1408–1415.

Bokhorst S, Bjerke JW, Davey MP, Taulavuori K, Taulavuori E, Laine K, Callaghan TV, Phoenix GK (2010) Impacts of extreme winter warming events on plant physiology in a sub-arctic heath community. Physiol. Plant. 140: 128–140.

Bokhorst S, Bjerke JW, Street LE, Callaghan TV, Phoenix GK (2011) Impacts of multiple extreme winter warming events on sub-arctic heathland: Phenology, reproduction, growth, and CO_2 flux responses. Glob. Change Biol. 17: 2817–2830.

Bokhorst S, Tommervik H, Callaghan TV, Phoenix GK, Bjerke JW (2012) Vegetation recovery following extreme winter warming events in the sub-arctic estimated using NDVI from remote sensing and handheld passive proximal sensors. Environ. Exp. Bot. 81: 18–25.

Bowler C, Vanmontagu M, Inze D (1992) Superoxide-dismutase and stress tolerance. Annu. Rev. Plant Physiol. Plant Mol. Biol. 43: 83–116.

Browse J, Xin ZG (2001) Temperature sensing and cold acclimation. Curr. Opin. Plant Biol. 4: 241–246.

Browse J, Lange BM (2004) Counting the cost of a cold-blooded life: Metabolomics of cold acclimation. Proc. Natl. Acad. Sci. USA 101: 14996–14997.

Callaghan TV, Jonasson S (1995) Arctic terrestrial ecosystems and environmental change. Philos. Trans. R. Soc. A-Math. Phys. Eng. Sci. 352: 259–276.

Castonguay Y, Dube M-P, Cloutier J, Bertrand A, Michaud R, Laberge S (2013) Molecular physiology and breeding at the crossroads of cold hardiness improvement. Physiol. Plant. 147: 64–74.

Chandler JW (2009) Auxin as compare in plant hormone crosstalk. Planta 231: 1–12.

Charra-Vaskou K, Mayr S (2011) The hydraulic conductivity of the xylem in conifer needles (*Picea abies* and *Pinus mugo*). J. Exp. Bot. 62: 4383–4390.

Chinnusamy V, Schumaker K, Zhu JK (2004) Molecular genetic perspectives on cross-talk and specificity in abiotic stress signalling in plants. J. Exp. Bot. 55: 225–236.

Cook D, Fowler S, Fiehn O, Thomashow MF (2004) A prominent role for the CBF cold response pathway in configuring the low-temperature metabolome of *Arabidopsis*. Proc. Natl. Acad. Sci. USA 101: 15243–15248.

Crawford R (2008) Plants at the margin: Ecological limits and climate change. Cambridge University Press, Cambridge.

Davey MP, Woodward FI, Quick WP (2009) Intraspecific variation in cold-temperature metabolic phenotypes of *Arabidopsis* lyrata ssp petraea. Metabolomics 5: 138–149.

Edgerton MD (2009) Increasing crop productivity to meet global needs for feed, food, and fuel. Plant Physiol. 149: 7–13.

Fowler S, Thomashow MF (2002) *Arabidopsis* transcriptome profiling indicates that multiple regulatory pathways are activated during cold acclimation in addition to the CBF cold response pathway. Plant Cell 14: 1675–1690.

Germino MJ, Smith WK, Resor AC (2002) Conifer seedling distribution and survival in an alpine-treeline ecotone. Plant Ecol. 162: 157–168.

Gilmour SJ, Zarka DG, Stockinger EJ, Salazar MP, Houghton JM, Thomashow MF (1998) Low temperature regulation of the *Arabidopsis* CBF family of AP2 transcriptional activators as an early step in cold-induced COR gene expression. Plant J. 16: 433–442.

Godfray C, Beddington JR, Crute IR, Haddad L, Lawrence D, Muir JF, Pretty J, Robinson S, Thomas SM, Toulmin C (2010) Food security: The challenge of feeding 9 billion people. Science 327: 812–818.

Gray GR, Chauvin LP, Sarhan F, Huner NPA (1997) Cold acclimation and freezing tolerance – A complex interaction of light and temperature. Plant Physiol. 114: 467–474.

Gu L, Hanson PJ, Mac Post W, Kaiser DP, Yang B, Nemani R, Pallardy SG, Meyers T (2008) The 2007 eastern US spring freezes: Increased cold damage in a warming world? Bioscience 58: 253–262.

Gusta LV, Wisniewski M (2013) Understanding plant cold hardiness: An opinion. Physiol. Plant. 147: 4–14.

Gusta LV, Trischuk R, Weiser CJ (2005) Plant cold acclimation: The role of abscisic acid. J. Plant Growth Regul. 24: 308–318.

Guy CL (1990) Cold acclimation and freezing stress tolerance – role of protein metabolism. Annu. Rev. Plant Physiol. Plant Mol. Biol. 41: 187–223.

Guy C, Kaplan F, Kopka J, Selbig J, Hincha DK (2008) Metabolomics of temperature stress. Physiol. Plant. 132: 220–235.

Hacker J, Neuner G (2008) Ice propagation in dehardened alpine plant species studied by infrared differential thermal analysis (IDTA). Arct. Antarct. Alp. Res. 40: 660–670.

Hacker J, Ladinig U, Wagner J, Neuner G (2011) Inflorescences of alpine cushion plants freeze autonomously and may survive subzero temperatures by supercooling. Plant Sci. 180: 149–156.

Hanin M, Brini F, Ebel C, Toda Y, Takeda S, Masmoudi K (2011) Plant dehydrins and stress tolerance: Versatile proteins for complex mechanisms. Plant Signal. Behav. 6: 1503–1509.

Hannah MA, Wiese D, Freund S, Fiehn O, Heyer AG, Hincha DK (2006) Natural genetic variation of freezing tolerance in *Arabidopsis*. Plant Physiol. 142: 98–112.

Harberd NP, Belfield E, Yasumura Y (2009) The angiosperm gibberellin-GID1-DELLA growth regulatory mechanism: How an "Inhibitor of an Inhibitor" enables flexible response to fluctuating environments. Plant Cell 21: 1328–1339.

Heidarvand L, Amiri RM (2010) What happens in plant molecular responses to cold stress? Acta Physiol. Plant. 32: 419–431.

Hirsh AG, Williams RJ, Meryman HT (1985) A novel method of natural cryoprotection - intracellular glass formation in deeply frozen *Populus*. Plant Physiol. 79: 41–56.

Huner NPA, Oquist G, Hurry VM, Krol M, Falk S, Griffith M (1993) Photosynthesis, photoinhibition and low temperature acclimation in cold tolerant plants. Photosynth. Res. 37: 19–39.

Inouye DW (2008) Effects of climate change on phenology, frost damage, and floral abundance of montane wildflowers. Ecology 89: 353–362.

Ishikawa M (1984) Deep supercooling in most tissues of wintering *Sasa senanensis* and its mechanism in leaf blade tissues. Plant Physiol. 75: 196–202.

Janska A, Marsik P, Zelenkova S, Ovesna J (2010) Cold stress and acclimation – what is important for metabolic adjustment? Plant Biol. 12: 395–405.

Juntilla O, Robberecht R (1999) Ecological aspects of cold-adapted plants with special emphasis on environmental control of cold hardening and dehardening. In: Margsin R, Schinner F (eds). Cold-adapted organisms – ecology, physiology, enzymology and molecular biology. Springer Verlag, Berlin, pp. 55–77.

Kalberer SR, Wisniewski M, Arora R (2006) Deacclimation and reacclimation of cold-hardy plants: Current understanding and emerging concepts. Plant Sci. 171: 3–16.

Kalberer S, Leyva-Estrada N, Krebs S, Arora R (2007) Cold hardiness of floral buds of deciduous azaleas: Dehardening, rehardening, and endodormancy in late winter. Hortscience 42: 974–974.

Kaplan F, Kopka J, Sung DY, Zhao W, Popp M, Porat R, Guy CL (2007) Transcript and metabolite profiling during cold acclimation of *Arabidopsis* reveals an intricate relationship of cold-regulated gene expression with modifications in metabolite content. Plant J. 50: 967–981.

Kappen, L (1993) Plant activity under snow and ice, with particular reference to lichens. Arctic, 46: 297–302.

Kasuga M, Liu Q, Miura S, Yamaguchi-Shinozaki K, Shinozaki K (1999) Improving plant drought, salt, and freezing tolerance by gene transfer of a single stress-inducible transcription factor. Nat. Biotechnol. 17: 287–291.

Knight MR, Knight H (2012) Low-temperature perception leading to gene expression and cold tolerance in higher plants. New Phytol. 195: 737–751.

Körner C (1999) Alpine plant life – functional plant ecology of high mountain ecosystems. 1st edn. Springer, Heidelberg.

Körner C (2003) Alpine plant life – functional plant ecology of high mountain ecosystems. 2nd edn. Springer, Heidelberg.

Körner C, Diemer M (1987) In situ photosynthetic responses to light, temperature and carbon dioxide in herbaceous plants from low and high altitudes. Func. Ecol. 1: 179–194.

Kosova K, Holkova L, Prasil IT, Prasilova P, Bradacova M, Vitamvas P, Capkova V (2008) Expression of dehydrin 5 during the development of frost tolerance in barley (*Hordeum vulgare*). J. Plant Physiol. 165: 1142–1151.

Krause GH (1994) Photoinhibition induced by low temperatures. In: Baker NR, Bowyer JR (eds). Photoinhibition of photosynthesis from molecular mechanisms to the field. BIOS Scientific Publishers, Oxford, pp. 331–348.

Kudo G, Nordenhall U, Molau U (1999) Effects of snowmelt timing on leaf traits, leaf production, and shoot growth of alpine plants: Comparisons along a snowmelt gradient in northern Sweden. Ecoscience 6: 439–450.

Larcher, W (1995) Photosynthesis as a tool for indicating temperature stress events. In: Schulze ED, Caldwell MM (eds). Ecophysiology of photosynthesis. Springer Verlag, New York, pp. 261–274.

Larcher W (2003) Physiological plant ecology: Ecophysiology and stress physiology of functional groups. Springer-Verlag, Berlin.

Larcher W, Kainmueller C, Wagner J (2010) Survival types of high mountain plants under extreme temperatures. Flora 205: 3–18.

Leinonen I, Repo T, Hanninen H (1997) Changing environmental effects on frost hardiness of *Scots pine* during dehardening. Ann. Bot. 79: 133–138.

Levitt J (1980) Responses of plants to environmental stress: Chilling, freezing and high temperature stresses. Academic Press, New York.

Loik ME, Huxman TE, Hamerlynck EP, Smith SD (2000) Low temperature tolerance and cold acclimation for seedlings of three Mojave Desert *Yucca* species exposed to elevated CO_2. J. Arid Environ. 46: 43–56.

Loveys BR, Egerton JJG, Ball MC (2006) Higher daytime leaf temperatures contribute to lower freeze tolerance under elevated CO_2. Plant Cell Environ. 29: 1077–1086.

Lütz C (2010) Cell physiology of plants growing in cold environments. Protoplasma 244: 53–73.

Lutze JL, Roden JS, Holly CJ, Wolfe J, Egerton JJG, Ball MC (1998) Elevated atmospheric CO_2 promotes frost damage in evergreen tree seedlings. Plant Cell Environ. 21: 631–635.

McKown R, Kuroki G, Warren G (1996) Cold responses of *Arabidopsis* mutants impaired in freezing tolerance. J. Exp. Bot. 47: 1919–1925.

Neuner G, Hacker J (2012) Ice formation and propagation in alpine plants. In: Lutz C (ed). Plant in alpine regions. Cell physiology of survival and adaptation strategies. Springer, Vienna, pp. 163–174.

Neuner G, Erler A, Ladinig U, Hacker J, Wagner J (2013) Frost resistance of reproductive tissues during various stages of development in high mountain plants. Physiol. Plant. 147: 88–100.

Nitsch JP (1957) Photoperiodism in woody plants. Proc. Am. Soc. Hortic. Sci. 70: 526–544.

Oakenfull RJ, Baxter R, Knight MR (2013) A C-Repeat Binding Factor Transcriptional Activator (CBF/DREB1) from European bilberry (*Vaccinium myrtillus*) induces freezing tolerance when expressed in *Arabidopsis thaliana*. PLoS One 8: e54119.

Obrist D, Arnone JA, Korner C (2001) In situ effects of elevated atmospheric CO_2 on leaf freezing resistance and carbohydrates in a native temperate grassland. Ann. Bot. 87: 839–844.

Pagter M, Arora R (2013) Winter survival and deacclimation of perennials under warming climate: Physiological perspectives. Physiol. Plant. 147: 75–87.

Pagter M, Williams M (2011) Frost dehardening and rehardening of *Hydrangea macrophylla* stems and buds. Hortscience 46: 1121–1126.

Pearce RS (2001) Plant freezing and damage. Ann. Bot. 87: 417–424.

Pearce RS, Ashworth EN (1992) Cell-shape and localization of ice in leaves of overwintering wheat during frost stress in the field. Planta 188: 324–331.

Penfield S (2008) Temperature perception and signal transduction in plants. New Phytol. 179: 615–628.

Pukacki PM, Kaminska-Rozek E (2013) Reactive species, antioxidants and cold tolerance during deacclimation of *Picea abies* populations. Acta Physiol. Plant. 35: 129–138.

Rahman A (2013) Auxin: a regulator of cold stress response. Physiol. Plant. 147: 28–35.

Raunkiaer C, Gilbert-Carter H, Fausbøll A, Tansley A (1934) The life forms of plants and statistical plant geography. The Clarendon Press, Oxford.

Rixen C, Dawes MA, Wipf S, Hagedorn F (2012) Evidence of enhanced freezing damage in treeline plants during six years of CO_2 enrichment and soil warming. Oikos 121: 1532–1543.

Roberts DWA (1984) The effect of light on development of the rosette growth habit of winter wheat. Can. J. Bot. 62: 818–822.

Sakai A, Larcher W (1987) Frost survival of plants: Responses and adaptation to freezing stress. Springer-Verlag, Berlin.

Sanghera GS, Wani SH, Hussain W, Singh NB (2011) Engineering cold stress tolerance in crop plants. Curr. Genomics 12: 30–43.

Semikhatova OA, Gerasimenko TV, Ivanova TI (1992) Photosynthesis, respiration, and growth of plants in the Soviet Arctic. In: Chapin FS, III, Jeffries RL, Reynolds JF, Shaver GR, Svoboda J (eds). Arctic ecosystems in a changing climate. Academic Press, San Diego, pp. 169–192.

Seo PJ, Lee A-K, Xiang F, Park C-M (2008) Molecular and functional profiling of *Arabidopsis* pathogenesis-related genes: Insights into their roles in salt response of seed germination. Plant Cell Physiol. 49: 334–344.

Shameer K, Ambika S, Varghese SM, Karaba N, Udayakumar M, Sowdhamini R (2009) STIFDB – *Arabidopsis* Stress Responsive Transcription Factor DataBase. Int. J. Plant Genomics 2009: 583429.

Sierra-Almeida A, Cavieres LA (2012) Summer freezing resistance of high-elevation plant species changes with ontogeny. Environ. Exp. Bot. 80: 10–15.

Srivastava LM (2002) Plant growth and development: Hormones and environment. Academic Press, San Diego.

Starr CR, Oberbauer SF (2003) Photosynthesis of arctic evergreens under snow: Implications for tundra ecosystem carbon balance. Ecology 84: 1415–1428.

Stitt M, Hurry V (2002) A plant for all seasons: Alterations in photosynthetic carbon metabolism during cold acclimation in *Arabidopsis*. Curr. Opin. Plant Biol. 5: 199–206.

Suojala T, Linden L (1997) Frost hardiness of *Philadelphus* and *Hydrangea* clones during ecodormancy. Acta Agric. Scand. Sect. B Soil Plant Sci. 47: 58–63.

Taschler D, Beikircher B, Neuner G (2004) Frost resistance and ice nucleation in leaves of five woody timberline species measured in situ during shoot expansion. Tree Physiol. 24: 331–337.

Taulavuori K, Taulavuori E, Niinimaa A, Laine K (2001) Acceleration of frost hardening in *Vaccinium* vitis-idaea by nitrogen fertilization. Oecologia 127: 321–323.

Taulavuori KMJ, Taulavuori EB, Skre O, Nilsen J, Igeland B, Laine KM (2004) Dehardening of mountain birch (*Betula pubescens* ssp *czerepanovii*) ecotypes at elevated winter temperatures. New Phytol. 162: 427–436.

Thomashow MF (1999) Plant cold acclimation: Freezing tolerance genes and regulatory mechanisms. Ann. Rev. Plant Physiol. Plant Mol. Biol. 50: 571–599.

Thomashow MF (2010) Molecular basis of plant cold acclimation: Insights gained from studying the CBF cold response pathway. Plant Physiol. 154: 571–577.

Tuteja N, Tiburcio AF, Fortes AM, Bartels D (2011) Plant abiotic stress. Introduction to PSB special issue. Plant Signal. Behav. 6: 173–174.

Usadel B, Blaesing OE, Gibon Y, Poree F, Hoehne M, Guenter M, Trethewey R, Kamlage B, Poorter H, Stitt M (2008) Multilevel genomic analysis of the response of transcripts, enzyme activities and metabolites in *Arabidopsis* rosettes to a progressive decrease of temperature in the non-freezing range. Plant Cell Environ. 31: 518–547.

Warren G, McKown R, Marin A, Teutonico R (1996) Isolation of mutations affecting the development of freezing tolerance in *Arabidopsis thaliana* (L) Heynh. Plant Physiol. 111: 1011–1019.

Welling A, Palva ET (2008) Involvement of CBF transcription factors in winter hardiness in birch. Plant Physiol. 147: 1199–1211.

Wisniewski, M, Lindow SE, Ashworth EN (1997) Observations of ice nucleation and propagation in plants using infrared video thermography. Plant Physiol. 113: 327–334.

Woldendorp G, Hill MJ, Doran R, Ball MC (2008) Frost in a future climate: Modelling interactive effects of warmer temperatures and rising atmospheric CO_2 on the incidence and severity of frost damage in a temperate evergreen (*Eucalyptus pauciflora*). Glob. Change Biol. 14: 294–308.

Woolgrove CE, Woodin SJ (1996) Ecophysiology of the snow-bed bryophyte Kiaeria starkei during snow-melt and uptake of nitrate from meltwater. Can J Bot 75: 1095–1103.

Wolfe J, Bryant G (1999) Freezing, drying, and/or vitrification of membrane-solute-water systems. Cryobiology 39: 103–129.

Xin ZG, Browse J (1998) eskimo1 mutants of *Arabidopsis* are constitutively freezing-tolerant. Proc. Natl. Acad. Sci. USA 95: 7799–7804.

Yan SP, Zhang QY, Tang ZC, Su WA, Sun WN (2006) Comparative proteomic analysis provides new insights into chilling stress responses in rice. Mol. Cell. Proteomics 5: 484–496.

Zhou MQ, Shen C, Wu LH, Tang KX, Lin J (2011) CBF-dependent signaling pathway: A key responder to low temperature stress in plants. Crit. Rev. Biotechnol. 31: 186–192.

3 Plant acclimation and adaptation to warm environments

Martijn van Zanten, Ralph Bours, Thijs L. Pons,
and Marcel C.G. Proveniers

3.1 Introduction

One of the most important abiotic environmental parameters directly affecting plant fitness is temperature. Although some plants are thermogenic and capable of maintaining a stable inflorescence temperature (Seymour 2001), most plants have to deal with the temperature imposed on them by the environment. Temperature has a major impact on the functioning and distribution of plant species, and the ability to adjust to temperature fluctuations within the ambient range and beyond (stress) is often limiting for their geographic distribution (Hoffmann 2002; Walther 2003; Sexton et al. 2009). Despite this, plant life thrives in almost every corner of our planet, including the harshest habitats imaginable such as hot deserts and cold arctic tundra. Temperature can change rapidly within a day and can fluctuate pronouncedly over seasons. For instance, in temperate continental climates, winters can be very cold and summers can be dry and hot. Moreover in most, if not all, climates, diurnal differences in day- and nighttime temperature occur (temperature cycles).

A variety of adaptations have evolved in plants to cope with the challenges imposed by high temperature. These adaptations can be part of the general developmental plan of the plant, such as thick wax layers in dessert plants and cactus spines that function in collecting dew water and reflecting sunlight. In contrast, plants adapted to cold environments are often dwarfed with small, thick, compact leaves and possess very specific adaptations such as alterations in membrane lipid composition and cryoprotective protein accumulation (Smallwood and Bowles 2002; Atkin et al. 2006a). Plant adaptations to cold environments are discussed in more detail in Chapter 2.

Besides constitutive adaptations to high temperature, many plant species also possess mechanisms to adjust development and physiology to control the impact of high temperature during their lifetime. Collectively this is called acclimation. Acclimation is important to optimize vital processes

Temperature and Plant Development, First Edition. Edited by Keara A. Franklin
and Philip A. Wigge.

such as photosynthesis under a given temperature regime and to maintain a favorable water balance. Recently, Athanasiou et al. (2010) defined two distinct types of acclimation: (a) developmental acclimation (the adjustment of development during plant growth) and (b) dynamic acclimation (adjustment of fully developed organs to a changing environment). Atkin et al. (2006b) propose that the strategy followed differs among species. They suggest that for short living species, developmental acclimation of new leaves is more efficient than dynamic acclimation of existing ones. Therefore, in annual species like the model plant *Arabidopsis thaliana*, the lifetime benefit of acclimation to heat stress would depend on the developmental rate by which new leaves and other organs are formed after the stress occurs.

Acclimation to high temperature and heat stress, together with their underlying molecular, physiological, and genetic networks, has, like many other traits, been most intensively studied using *A. thaliana*. Using genome-wide analyses, it was shown that amino acid-changing small nucleotide polymorphisms (SNPs) in *Arabidopsis* were significantly enriched among loci that correlated with climate variables, including temperature parameters (Hancock et al. 2011). This indicates that different accessions have adapted to local temperature conditions by means of natural selection. For example, naturally occurring intron-specific sequence variation in *Δ^1-pyrroline-5-carboxylate synthetase1* (*P5CS1*) was shown to influence biosynthesis of osmoprotective proline, by affecting alternative splicing (Kesari et al. 2012). Signs for positive selection of P5CS1 were identified, and correlation analysis with various climate parameters indicated that diurnal temperature range at the collection site explains a large fraction of the observed sequence variation.

The dose–response kinetics of many temperature-dependent growth responses in *Arabidopsis* is very similar for independent processes within a given genotype (Parent and Tardieu 2012). These include leaf expansion rate, germination rate, cell elongation, and seedling elongation (Granier et al. 2002; Orbović and Poff 2007; Parent et al. 2010; Vasseur et al. 2011; Watt and Bloomberg 2012). Together, these examples suggest that the developmental and physiological processes involved in acclimation to high temperature have a coordinated temperature-dependent response within a given genotype/accession. The optimum temperature may be close to the temperature the accession encounters in the natural environment during the growth season. Indeed, frequencies of alleles (SNPs) that affect fitness in a given climate were enriched on sites where this environment prevails. Alleles associated with survival and fitness traits (silique number) were particularly constrained by temperature variables (Fournier-Level et al. 2011). Empirical evidence that the candidate genes carrying these SNPs do indeed significantly contribute to enhanced fitness under the local temperature regime is, however, still lacking.

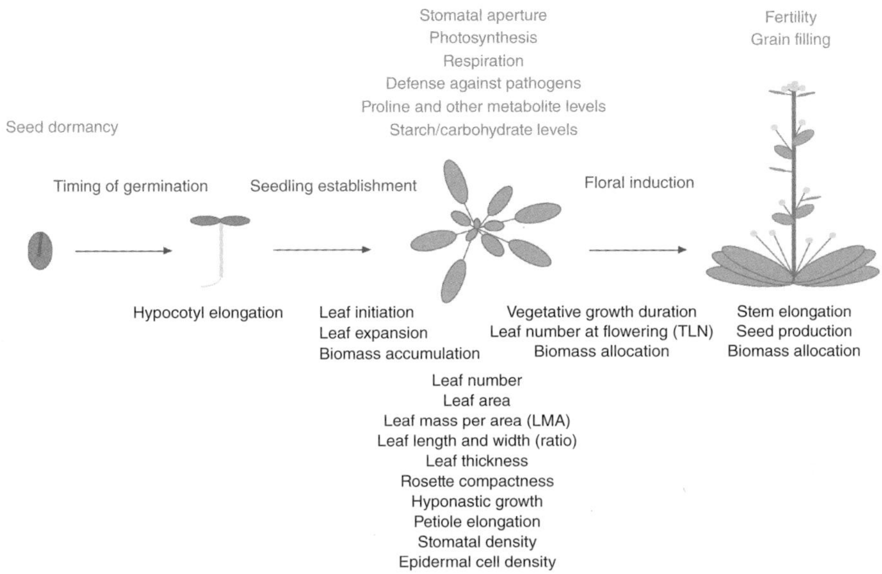

Figure 3.1 Schematic overview of morphological/developmental (*purple*), physiological (*blue*), and phase transition (*red*) traits affected by high temperature throughout the life history of the flowering rosette plant *A. thaliana*. For color detail, please see color plate section.

These examples underline that *Arabidopsis* is an excellent model system to study the genetic basis of adaptations and the mechanisms of acclimation that have evolved to withstand high temperature. In this chapter, we highlight some adaptations and acclimation responses to high temperature, with the focus on *Arabidopsis* (summarized in Figure 3.1). Elongation growth, leaf movement, and changes in flowering time are among the best-studied responses to high temperature and heat stress in this species. These responses will be discussed, including variation therein as adaptation to local climate, along with general effects on plant development, architecture, photosynthesis, respiration, and biomass reallocation.

3.2 Implications of high temperature for agriculture and natural ecosystems

Current global warming has major impact on the growth, survival, and phenology of individual plants and communities (Walther 2003; Nicotra et al. 2010). The average first flowering date in Great Britain, for example, advanced significantly in the 1990s compared to foregoing decades (Fitter and Fitter 2002), and mountainous species have tended to shift to higher altitudes (Pauli et al. 2012). It has been predicted that, owing to the effects of global warming, up to half of the European species will become vulnerable or extinct by 2080 (Thuiller et al. 2007). Already, plant species which are unable to adjust their flowering time in response to changing temperature have disappeared or were greatly reduced in abundance on certain natural sites (Willis et al. 2008). Several of the traits that are affected by high temperature are directly or indirectly relevant for agriculture. These include seed dormancy, timing of germination, stem elongation, leaf expansion, flowering, grain filling capacity, and fertility (Probert 2000; Schmuths et al. 2006; Parent and Tardieu 2012). High-temperature stress through global warming is therefore a significant threat to both agricultural productivity and future food security (Battisti and Naylor 2009). In combination with drought, heat is already among the most damaging abiotic factors affecting crop plants, with estimated losses to US agriculture alone exceeding 120 billion dollars in the period 1980–2004 (Mittler 2006; Barnabás et al. 2008).

Despite centuries of breeding, the effects of short-term changes in temperature on the development and performance of important staple crops, such as barley, rice, and maize, have not been majorly affected. This is mainly due to a lack of genetic variation in the germplasm (Parent and Tardieu 2012). Breeding for more heat-tolerant crop varieties therefore needs to focus on improvement of composite traits that contribute to adaptation and acclimation to high temperature. Understanding the underlying molecular and physiological mechanisms is essential to secure and improve future food production and to predict and counteract effects of global warming.

3.3 Temperature perception and signaling pathways

Temperature affects growth and development throughout the plant's life history and controls the timing of major phase transitions such as germination and flowering (Figure 3.1). Although the mechanisms of how plants perceive temperature remained elusive for a long time, different temperature perception and signaling pathways are now being unravelled. This topic is discussed in several reviews (e.g., Kotak et al. 2007; Penfield 2008; McClung and Davis 2010; Mittler et al. 2012) and is the focus of Chapter 1.

At the cellular and molecular level, heat stress interferes with many vital processes as it directly affects stability of proteins, membranes, RNA species, and the structure of the cytoskeleton. To counter the effects of heat stress, major changes in the metabolome, transcriptome, and proteome are induced (reviewed in Kotak et al. 2007; Penfield 2008; Mittler et al. 2012). Increased temperature and heat shock alter membrane properties and activate calcium channels. The resulting calcium flux induces a signal transduction cascade involving complex regulatory networks of heat shock factor (HSF) transcription factors, which subsequently induces molecular chaperones and heat shock proteins (HSP). These orchestrate acclimation at the whole-plant level. Accordingly, it was found by genome-wide transcriptional analysis that global gene expression levels obtained from *Arabidopsis* plants from the field strongly correlated with the temperature (and precipitation) encountered (Richards et al. 2012).

Recent groundbreaking work by Kumar and Wigge (2010) elegantly demonstrated a direct interaction between temperature and chromatin-level regulation of gene expression. The authors showed that the specific histone H2A variant (H2A.Z) has a direct thermosensory function. In short, the occupancy of H2A.Z at promoters of temperature-regulated genes decreases when temperature rises, altering accessibility to transcription machinery. Mutants in *ACTIN-RELATED PROTEIN6* (*ARP6*), which are disturbed in H2A.Z positioning, show exaggerated acclimation responses to high-temperature stress. In these lines, high-temperature-induced hypocotyl and petiole elongation responses were enhanced compared to wild type (Kumar and Wigge 2010). High temperatures also affect levels of plant DNA methylation (Pecinka et al. 2010; Tittel-Elmer et al. 2010), providing further evidence for the involvement of epigenetic processes. Specific DNA methylation states can be transmitted to next generations. Transgenerational inherited epigenetic markers are therefore potent signals by which a mother plant can pass on information about the temperature environment it experienced to its offspring. In one interesting example, Whittle et al. (2009) demonstrated the occurrence of transgenerational memory of ancestral high temperatures in reproductive output (seed production per individual). In this case, the reproductive output of *Arabidopsis* plants increased over fivefold in F2 progeny of homozygous parents and F1's that had been subjected to mild heat stress (30°C). These effects persisted through the F3 progeny, even though the F2 generations did not experience a heat treatment. Although the exact mechanistic basis for this phenomenon is yet unclear, it is likely to be based on stable heritable epigenetic modifications.

3.4 Photosynthesis

The process of photosynthesis supplies the metabolic energy and carbon necessary for plant growth and maintenance. Carbon gain through photosynthesis (CO_2 assimilation) as well as carbon loss through respiration depends on temperature (Berry and Björkman 1980). High-temperature effects on photosynthesis in C_3 plants can be roughly divided in three temperature ranges: (a) warm temperatures that cause a gradual decrease in net photosynthesis (A_n) above the optimum temperature, whereas photosynthetic electron transport rate (ETR) still increases; (b) moderate heat stress that causes a reversible decrease of ETR and triggers mechanisms that protect the photosynthetic apparatus; and (c) severe heat stress that causes irreversible damage to the photosynthetic membranes (Sharkey and Zhang 2010).

The temperature ranges over which these processes occur vary with adaptation to climate. The optimum temperature for A_n at light saturation and atmospheric CO_2 concentration can vary between 6°C in arctic lichens and 40°C in hot desert perennials (Lambers et al. 2008). For temperate plants, the optimum temperature for the light-saturated rate of A_n is generally around 30°C. A higher growth temperature, however, typically results in a higher optimum temperature (Berry and Björkman 1980). In this way, the optimum temperature for photosynthesis is subject to acclimation. The decrease of A_n in warm conditions is the result of the increase of the oxygenation reaction of Rubisco at the expense of carboxylation. This is partly the result of a stronger increase of the K_m value for the carboxylation reaction of Rubisco, compared to the oxygenation reaction, which results in a lower specificity for CO_2. The solubility of CO_2 decreases faster with increasing temperature compared to O_2, resulting in a lower CO_2/O_2 ratio in the chloroplast stroma. The increased oxygenation of ribulose-1, 5-bisphosphate in warm conditions leads to increased photorespiration. These added effects result in a substantial decrease of A_n, although ETR is still increasing. Photorespiration is suppressed at high CO_2 concentrations and also in C_4 plants. In these circumstances, A_n is proportional to ETR, and thus the optimum temperature for A_n is higher (Sage and Kubien 2007).

Light conditions also influence the optimum temperature for A_n. For example, *Arabidopsis* grown at an irradiance of 300 μmol m^{-2} s^{-1} had an optimum temperature of about 22°C. This was higher than 27°C at light saturation. For plants grown at 50 μmol m^{-2} s^{-1}, the optimum temperature was lower than 10°C (Pons 2012). This can be explained by constant photon absorption with variation in temperature and thus a constant ETR across a broad temperature range in light-limited conditions. As the fraction of photorespiration increases with temperature, A_n decreases across the range where ETR is constant, causing a low optimum temperature in low light (Berry and Björkman 1980). The optimum temperature for A_n in light-limited growth room conditions is therefore typically much lower than the optimum measured at light saturation.

The insensitivity of photon absorption to temperature and the increasing enzyme activity at high temperatures cause an imbalance between light harvesting and photochemistry in cool-grown plants transferred to warm conditions. Photosynthetic acclimation to warm temperatures reestablishes the balance (Hikosaka 1997). The investment in photosynthetic proteins associated with photosynthetic capacity decreases relative to pigment–protein complexes at high temperatures. For example, Rubisco per unit chlorophyll was 20% lower in warm- compared to cool-grown *Arabidopsis* (Pons 2012). Pigment content can also be somewhat lower in warm conditions

Figure 3.2 Morphological acclimation of *A. thaliana* to increased ambient temperature and high-temperature stress. (A, B) Rosette morphology of 30-day-old plants grown continuously at (A) 22°C or (B) 27°C, under LD conditions. Note the open rosette structure in warm-grown plants with elongated petioles, erect leaves, early flowering (elongated inflorescence), lower pigmentation (chlorophyll), and thin leaves (B) compared to those in panel (A). (C, D) Heat-induced hyponastic growth phenotype of (C) Col-0 plant grown at 20°C and (D) the same plant treated for 6 h with high temperature (38°C). Note the orange-painted dots, used to facilitate leaf angle quantification. (E, F) High-temperature-induced hypocotyl elongation. Seven-day-old Col-0 seedling grown at (E) 22°C and (F) 27°C. Note that high temperature induces not only hypocotyl elongation but also hyponastic growth of cotyledons. Scale bar is 5 mm. (G, H) Effect of DIF treatment on whole rosette morphology. (G) Plants grown under +DIF (22°C day, 12°C night) and (H) −DIF (12°C day, 22°C night). Note the typical compact rosette phenotype of plants grown under −DIF. For color detail, please see color plate section.

(Figure 3.2A, B). The combined effect of reduced photosynthetic capacity per unit chlorophyll and chlorophyll per unit leaf area reduces the size of the photosynthetic apparatus per unit leaf area. Fewer chloroplasts require less space in the mesophyll (Terashima et al. 2011), making warm-grown leaves thinner than cool-grown ones. This results in a similar A_n in *Arabidopsis* when expressed per unit leaf area (Armstrong et al. 2006a) but a higher rate when expressed per unit leaf dry mass. Cold-tolerant species have a greater capability for acclimation to growth temperature compared to cold-sensitive species that originate from warmer climates (Yamori et al. 2009). Photosynthetic acclimation to warm and cool conditions was, however, rather similar in two *Arabidopsis* accessions from contrasting climates (subtropical Cape Verde Islands and boreal Scandinavia) (Pons 2012). This contrasts with other species, where intraspecific variation of photosynthetic traits with climate is found (Flood et al. 2011).

Plants can tolerate moderate heat stress for short periods. For instance, in *Arabidopsis*, a reversible decrease in the efficiency of photosystem II (ΦPSII) occurred with a 30-min short exposure to 40°C, whereas an irreversible decrease occurred after exposure to 44°C (Sharkey and Zhang 2010). Under moderate heat stress, regulatory mechanisms are induced that limit damage to the thylakoid membrane (Sharkey and Zhang 2010). PSII is most sensitive to heat, with destruction of the D1 protein as the first sign of damage. This can be measured by chlorophyll fluorescence as a decrease in ΦPSII. An increase in thylakoid membrane fluidity is likely to be involved, which interferes with protein interaction and D1 repair. PSI is more stable at high temperature. Phosphorylation of the light-harvesting complex associated with PSII (LHCII) causes its partial migration to PSI. Excitation of PSI then increases relative to PSII, which increases cyclic electron transport at the expense of linear transport. This scenario results in increased proton transport into the thylakoid lumen. Simultaneously, the activity of Rubisco is downregulated as a result of the degradation of Rubisco activase (Crafts-Brandner and Salvucci 2000). This limits the consumption of ATP and thus maintains a low pH in the thylakoid lumen to protect the thylakoid, in part, by the induction of zeaxanthin formation in the xanthophyll cycle. Expression of a more heat-stable Rubisco activase in *Arabidopsis* improved photosynthetic performance and growth at high temperature (Kurek et al. 2007). Hence, degradation of Rubisco activase may also have a direct role in reducing photosynthesis under heat stress.

Another mechanism for protection of the thylakoid membrane that operates in many species is isoprene formation. The protective action of isoprene at high temperature was elegantly demonstrated by heterologous expression of the *Platanus* (*Platanus orientalis*) isoprene synthase (*IspS*) in *Arabidopsis* (a non-isoprene emitter) and by inhibiting isoprene synthesis in *Platanus* (Velikova et al. 2011). The small hydrocarbon partitions to the membrane, but the mechanism of its protective action is yet unknown. The compound is volatile, and in warm conditions, large quantities can escape to the atmosphere (Monson et al. 2012). Here, it is involved in photochemical reactions resulting in ozone formation above forests (Monson and Holland 2001).

3.5 Respiration and carbon balance

Increasing temperature typically increases respiration rate in a more or less exponential manner. The stimulating effect of short-term exposures continues until high temperatures are reached (Atkin and Tjoelker 2003). This contrasts with net photosynthesis that typically has a much lower optimum temperature, particularly at limiting irradiance, as discussed (see Section 3.4). This is not consistent with the rather constant ratio between whole-plant daily respiration and photosynthesis (*R*/*A*), which shows a tendency of homeostasis across a broad temperature range in several species (Gifford 1995; Atkin et al. 2006b). The short-term effect of increasing temperature is the result of increasing activity of enzymes involved in ATP generation as well as its utilization. In the moderate to warm temperature range, respiration is limited by demand for ATP that is utilized in biosynthesis and other energy-demanding processes (Atkin and Tjoelker 2003). As a consequence, respiration decreases after a prolonged exposure to warm temperatures to levels that can be met by the supply of substrate for growth. Downregulation of growth rate is likely to occur before carbohydrates are exhausted, with a concomitant decrease in respiration. Sugar sensing has been suggested to perform a central role in this regulation (Smith and Stitt 2007). When compared to cold conditions, warm-grown *Arabidopsis* leaves showed a lower respiratory capacity (Armstrong et al. 2007). This resulted in a similar *R*/*A* between the two temperature conditions (Gorsuch et al. 2010). The adjustment of respiration rate to the temperature regime, involving structural changes, is referred to as acclimation of respiration.

When temperatures exceed the range that normally occurs in the plant's native habitat, *R*/*A* can increase dramatically with negative consequences for the carbon balance of these plants. For instance, the whole-plant *R*/*A* of the alpine species *Plantago euryphylla* was 0.28 at 13°C but increased to 0.39 in 27°C-grown plants and further to 0.60 when these plants were measured at 34°C. This contrasts with the temperate lowland species *Plantago major* that had a homeostatic *R*/*A* value across that temperature range of *ca.* 0.26 (Atkin et al. 2007). The increase of *R*/*A* at warm temperature in plants that cannot maintain a homeostatic value can result from increasing respiration, decreasing photosynthesis, or both (Loveys et al. 2002). As *Arabidopsis* is a fast-growing temperate species that shows substantial photosynthetic acclimation to temperature (Stitt and Hurry 2002; Pons 2012), its *R*/*A* value is likely to display a high degree of homeostasis across a broad temperature range (Yamori et al. 2009). Warm-temperature effects on the carbon balance of *Arabidopsis* have not, however, been systematically investigated.

3.6 Growth and allocation of biomass

Leaves of plants grown at warm temperatures are often thinner compared to cool-grown leaves, which is associated with lower photosynthetic capacity per unit area. *Arabidopsis* displays this type of acclimation to high temperature (Figure 3.2A, B; Vile et al. 2012). The leaf cells are also larger in warm-grown *Arabidopsis*, making the dry matter density of the tissue lower. Together, this results in reduced dry mass per unit leaf area (LMA) (Armstrong et al. 2006b; Gorsuch et al. 2010; Pyl et al. 2012). The largest and most consistent changes in LMA are, however, generally found across the cool temperature range (Loveys et al. 2002; Poorter et al. 2009).

Allocation of biomass between roots and leaves is also affected by temperature. Warm-grown plants generally have a lower fraction of biomass in roots (RMF) in favor of the fraction in leaves (LMF), when compared to their cool-grown counterparts (Poorter et al. 2012). As with LMA, the effects of temperature are generally less consistent for the warmer temperature ranges. The resistance for water uptake by roots is much reduced at warmer temperatures. Hence, a smaller root system is required to compensate for enhancement in water loss by transpiration in warm conditions (Lambers et al. 2008).

The relative growth rate (RGR, dry mass increase per dry mass present) of several plant species is maximum at moderate temperatures in growth chambers. Above the optimum temperature, the net assimilation rate (NAR, dry mass increase per unit leaf area) has been shown to decrease strongly in these plants (Loveys et al. 2002), reflecting the effect of warm temperatures on respiration and photosynthesis as discussed previously. The negative warm-temperature effects on RGR ($RGR = NAR \times LMF \times 1/LMA$) were thus dominated by the negative effects on NAR, which was minimally counteracted by a small decrease in LMA and increase in LMF. The representativeness of such growth chamber experiments for plants growing in natural environments and association of temperature effects on growth rate with climatic origin do, however, require further investigation.

3.7 Architectural changes in response to high temperature

In addition to the previously described effects on plant growth and biomass allocation, high temperatures result in suite of architectural phenotypes. *Arabidopsis* plants grown under a high-temperature regime display dramatic hypocotyl (Figure 3.2E, F) and petiole elongation (Figure 3.2A, B), compared to control plants grown at a cooler temperature (Gray et al. 1998; Koini et al. 2009; Crawford et al. 2012). Warm-grown adult rosette plants also display erect (inclined) leaves (Figure 3.2C, D). This phenotype can be either constitutive or induced by a sudden temperature rise. Active leaf repositioning is termed hyponastic growth.

3.7.1 Heat-induced hyponastic growth in Arabidopsis *and hormonal and light control*

Hyponastic growth is among the first visual responses to a sudden increase in temperature in several rosette species such as *A. thaliana* (Figure 3.2C–F) (Millenaar et al. 2005; Koini et al. 2009; Patel et al. 2013; van Zanten et al. 2009, 2010; Vasseur et al. 2011) and *Salvia broussonetii* (Abreu and Munné-Bosch 2008). Within an hour of high-temperature treatment, hyponastic growth is induced in vegetative *Arabidopsis* plants (van Zanten et al. 2009, 2010). It can be clearly seen in Figure 3.2E, F that very young seedlings are also hyponastic when grown at high temperature.

Heat-induced hyponastic growth follows the Arrhenius-type dose–response curve, typical for temperature-dependent biological processes. A mild increase in temperature leads to pronounced leaf angle readjustment. When *Arabidopsis* accessions were pre-grown at 20°C, the maximum difference in leaf angle was reached at 38°C (van Zanten et al. 2009). Besides high-temperature treatment, application of the phytohormone ethylene, which is a mimic for submergence (Millenaar et al. 2005, 2009; Benschop et al. 2007); proximity of neighboring vegetation; and low light intensities are able to induce hyponastic growth (Hangarter 1997; Millenaar et al. 2005, 2009; Pierik et al. 2005; Mullen et al. 2006; Vasseur et al. 2011). The cellular basis and hormonal control of hyponastic growth are relatively well understood (Polko et al. 2012). The response is driven by specific cell elongation in a defined, approximately 2-mm-long zone of proximally located abaxial epidermal cells and, at least the response induced by ethylene, requires reorientation of cortical microtubules (Polko et al. 2012). Since the kinetics of the hyponastic growth responses induced by the different abiotic factors are strikingly similar, it is plausible that the same mechanisms also apply to the high-temperature response.

Genetic and pharmacological evidence has shown that heat (38°C)-induced hyponasty is inhibited by ethylene (van Zanten et al. 2009). Furthermore, auxin and polar auxin transport (Koini et al. 2009; van Zanten et al. 2009), gibberellins (GA) (Peña-Castro et al. 2011), and abscisic acid (ABA) are involved in the response (van Zanten et al. 2009, 2010). A role for membrane repair processes in the determination of leaf movement/angle has also been suggested by studies of *heat-intolerant1* (*hit1*) mutants. These plants are disturbed in retrograde trafficking of vesicles involved in thermal adaptation of the plasma membrane (Wang et al. 2011). Although the *hit1* mutant is visually similar to wild-type plants at 22°C, transfer to high temperature (37°C) revealed impairment of leaf movement (Wu et al. 2000).

Extensive natural variation has been found for hyponastic growth among natural accessions (van Zanten et al. 2009, 2010). Some have even been shown to display epinasty (the downward curvature of leaves) in response to heat treatment. Using a subset of these accessions, representing the latitudinal biogeographic distribution of the species, it was found that the hyponastic response correlates with diurnal temperature variation at the site of collection (van Zanten et al. 2009). These observations strengthen the notion by Hopkins et al. (2008) that leaf movement is an adaptation to cope with warm environments.

Several developmental responses to changing light and changing temperature conditions are similar, and the molecular signaling pathways are tightly connected (Penfield 2008; Franklin 2009). Interactions between light and temperature signaling pathways are discussed in Chapter 5. The hyponastic growth responses of natural accessions to heat and low light intensity treatment strongly correlate (van Zanten et al. 2009, 2010). Moreover, the elongated petiole and small leaf phenotype of *Arabidopsis* plants grown under high temperature strongly resemble those of the shade-avoidance response, induced by proximity of neighbors (Franklin 2009; Patel and Franklin 2009). Maintenance of biomass under high temperatures (Foreman et al. 2011) and temperature-dependent internode elongation (Mazzella et al. 2000) both require action of the red and far-red light photoreceptor phytochrome B (phyB) and the blue light photoreceptor cryptochrome1 (cry1). Intriguingly, phyA, phyB, cry1, and cry2 are required for proper induction of heat-induced hyponastic growth (van Zanten et al. 2009). phyB is a negative regulator of the response amplitude to heat in both the Col-0 and Landsberg *erecta* (L*er*) accessions, whereas loss-of-function *phyA* rescues this effect in L*er*. Both cryptochromes are redundantly required for a fast induction of the response. Temperature effects on flowering time also depend on light perception and are discussed later in this chapter.

3.7.2 High-temperature-induced hypocotyl elongation in Arabidopsis

Growth of the embryonic stem (hypocotyl) by means of rapid cell elongation has been frequently used as a bioassay for high-temperature responsiveness. High-temperature-mediated hypocotyl elongation has been studied in a diverse range of species, including *Arabidopsis* (Gray et al. 1998) (Figure 3.2E, F), white mustard (*Sinapis alba*) (Wall and Johnson 1982), kenaf (*Hibiscus cannabinus*) (Carberry and Albrecht 1990), and lettuce (*Lactuca sativa*) (Stoddart et al. 1978). Supraoptimal high temperature (above 38°C) can, however, inhibit elongation growth in *Arabidopsis* (Hong and Vierling 2000).

In *Arabidopsis*, a shift from 20°C to 29°C results in increased levels of bioactive auxin (free indole-3-acetic acid (IAA)) and changes in gene expression of auxin-controlled genes (Gray et al. 1998). The involvement of auxin in hypocotyl elongation is supported by observations showing that high-temperature-induced hypocotyl elongation is reduced in mutant seedlings defective in auxin biosynthesis, responsiveness, and transport. These include mutants deficient in the temperature-responsive auxin biosynthesis gene *TRYPTOPHAN AMINOTRANSFERASE OF ARABIDOPSIS/TRANSPORT INHIBITOR RESPONSE 2 (TAA1/TIR2)* (Yamada et al. 2009) and the auxin receptors TIR1 and TIR3 (Gray et al. 1998). In contrast to auxin, genetic analysis has suggested that ethylene and ABA are not required for high-temperature-dependent hypocotyl elongation *per se*. Mutants deficient in a gene required for brassinosteroid biosynthesis, however, showed a moderately reduced response (Gray et al. 1998). Recent work by Stavang et al. (2009) has shown that the hormone gibberellic acid (GA) is important in the early phase of high-temperature-induced hypocotyl elongation, independently of auxin. Transcript levels of enzymes involved in GA biosynthesis were shown to increase with high-temperature treatment, while the accumulation of GA-regulated DELLA growth repressors decreased.

3.7.3 PIF4 as central regulator of high-temperature acclimation in Arabidopsis

Koini et al. (2009) showed that high-temperature-mediated hypocotyl and petiole elongation, as well as hyponastic growth, requires the basic helix–loop–helix (bHLH) transcription factor PHYTOCHROME-INTERACTING FACTOR 4 (PIF4). No roles were identified for other PIF family members, suggesting that PIF4 operates as a specific controller of high-temperature acclimation (Koini et al. 2009; Stavang et al. 2009). The role of PIF4 in temperature-mediated flowering is described in a subsequent section.

PIF4 is a positive regulator of cell elongation and was originally identified as a repressor of shade-avoidance responses (Huq and Quail 2002). Under standard conditions, phyB translocates to the nucleus where it binds and phosphorylates PIF4. This results in proteasomal degradation of the transcription factor and repression of PIF4-mediated elongation growth (Huq and Quail 2002; Lorrain et al. 2008). In the presence of neighboring vegetation, phyB is excluded from the nucleus, suppression of PIF4 is relieved, and the expression of genes involved in elongation growth is activated. High-temperature-induced hypocotyl elongation was unaltered in *phyBDE* triple (Koini et al. 2009) and *della*-quintuple null mutants (Koini et al. 2009; Stavang et al. 2009). PIF4-mediated responses therefore appear independent of DELLA and PHYTOCHROME action. Moreover, *HEAT*

SHOCK PROTEIN 70 (*HSP70*) transcription was unaltered in *pif4* mutants (Franklin et al. 2011). This gene is regarded a cellular thermometer since *HSP70* expression increases linearly with temperature (Kumar and Wigge 2010). It was therefore concluded that PIF4 does not affect general temperature sensing, but has a rather specific function in adjusting plant architecture and development to high temperature.

A transient increase in *PIF4* transcript abundance was observed in *Arabidopsis* hypocotyls transferred from 22°C to 29°C (Koini et al. 2009; Stavang et al. 2009; Sun et al. 2012). Two independent research groups have recently shown that PIF4 directly stimulates auxin biosynthesis by temperature-dependent binding to the following promoters of enzymes in the tryptophan (Trp)-dependent auxin biosynthesis pathway: *YUCCA8* (Sun et al. 2012), *TAA1/TAA1-RELATED* (*TAA1/TAR*), and the cytochrome P450 member *CYP79B2* (Franklin et al. 2011). Moreover, high-temperature-induced increases in free IAA levels were abolished in *pif4* mutants (Franklin et al. 2011; Sun et al. 2012), and hypocotyl elongation could be rescued by application of synthetic auxins (Franklin et al. 2011). Increased levels of auxin trigger proteasomal degradation of the *AUX/IAA* auxin response repressors. Gain-of-function mutations in these factors are typically auxin insensitive. Consistent with this notion, the *shy2/iaa3* gain-of-function mutant displays a shorter hypocotyl phenotype upon high-temperature treatment (Sun et al. 2012). Franklin et al. (2011) further showed that expression of SMALL AUXIN UP RNA 19–24 (*SAUR19–24*) genes is induced in a PIF4-dependent manner in high-temperature-treated hypocotyls. Intriguingly, 35S::GFP-*SAUR19* overexpression restored high-temperature-mediated hypocotyl elongation to *pif4* mutants, suggesting a major role for SAURS in driving growth (Franklin et al. 2011). Together, these observations suggest that PIF4 controls high-temperature-induced elongation growth by elevating free IAA levels via direct control of key enzymes involved in auxin biosynthesis and signaling.

3.8 Hormonal regulation of thermotolerance

In addition to modulating architectural changes, hormones have a role in conferring thermotolerance. Salicylic acid (SA), ethylene, and ABA have been shown to be involved in the protection against and repair of oxidative damage in *Arabidopsis* following heating. These hormones therefore contribute to plant survival under heat stress (Larkindale and Knight 2002). In later studies, hormone mutants were assayed for thermotolerance defects in seed germination, hypocotyl elongation, root growth, and seedling survival. It was found that ABA and SA stimulate acquired thermotolerance (the ability to withstand lethal temperatures if preceded by a short period of mild increase in temperature), whereas ethylene is seemingly involved more specifically in basal thermotolerance (Larkindale et al. 2005). SA and jasmonic acid (JA) cross talk has also been implicated in basal thermotolerance and presumably modulates ethylene production to protect against heat-mediated cell death (Clarke et al. 2009). In accordance with a role in protection against the effects of heat stress, pretreatment with methyl-JA (MeJA) and SA suppressed the hyponastic growth response to high-temperature treatment (van Zanten et al. 2012).

3.9 Functional implications of plant architectural changes to high temperature

The angle of leaves directionally to the sun determines the amount of light intercepted by the plant and consequently the flux of radiation-associated heat per unit leaf area. Leaf angle variations and leaf movements have therefore been suggested to be an adaptation to optimize photosynthesis, by preventing photoinhibition and controlling heat flux (King 1997; Falster and Westoby 2003; Hopkins et al. 2008; Koini et al. 2009; van Zanten et al. 2010; Crawford et al. 2012). Indeed, under well-watered conditions, paraheliotropism (light-avoiding movements) keeps leaf temperature and photosynthesis constant and close to optimal in *S. broussonetii* (Abreu and Munné-Bosch 2008) and several legumes and bean (*Phaseolus acutifolius* and *Phaseolus vulgaris*) (Fu and Ehleringer 1989; Yu and Berg 1994). A role for the photosynthetic apparatus in determining leaf angle in *Arabidopsis* has further been implicated by experiments in which the PSII electron transport chain was inhibited by 3-(3,4-dichlorophenyl)-1,1-dimethylurea (DCMU) (Millenaar et al. 2009).

As an example of how leaf angle and size interact in nature, Hopkins et al. (2008) showed that *Arabidopsis* accessions from Southern latitudes had generally more erect leaves than those from Northern latitudes. The authors propose that high leaf angles at low latitudes optimize light capture early in the day and during winter. During midday and summer, the same adaptations may prevent over-irradiation, supraoptimal heat flux, and water loss by avoiding direct vertical light on leaf lamina. These observations fit predictions for optimal photosynthetic ability across light intensity gradients over the globe (Falster and Westoby 2003) and suggest that natural selection occurred on leaf angle in *A. thaliana* to maximize photosynthesis under the environmental conditions of the native growth site. Leaf lengths are also correlated significantly with latitude (Hopkins et al. 2008). Plants from lower (Southern) latitudes were shown to have proportionally smaller leaf blades than those from high latitudes and have on average more elongated petioles.

It can be concluded from the earlier discussion that *Arabidopsis* plants grown under high temperatures (Koini et al. 2009; Vasseur et al. 2011; Crawford et al. 2012; Vile et al. 2012) or collected from a natural environment with prevalent high temperature (Southern latitudes) (Hopkins et al. 2008) are typically characterized by an open rosette architecture with few, small, thin erect leaves and elongated hypocotyls and petioles (see previous section and Figure 3.2A–F). Heat dissipation by convection is inversely proportional to the thickness of the boundary layer surrounding the leaves. Boundary layer thickness generally correlates to leaf size. It has therefore been suggested that the relatively small, thin leaves of warm-grown *Arabidopsis* plants may be more effective at dissipating heat (Crawford et al. 2012). In addition, the combination of elongated petioles, elongated hypocotyls, and the high leaf angle moves leaves away from the soil surface, increasing exposure to cooling breeze (Gray et al. 1998; Crawford et al. 2012). Empirical evidence has shown that plants pre-grown at warm temperature (showing an open rosette structure and hyponastic leaves) displayed a lower leaf temperature at 28°C than compact plants pre-grown at 22°C (Crawford et al. 2012). In conclusion, the suite of architectural and morphological traits observed in warm-grown *Arabidopsis* plants likely enhances leaf cooling capacity.

3.10 Interactions between drought and high temperature

Despite improving leaf cooling capacity, high-temperature-mediated architectural adaptations (Figure 3.2A–F) render plants more susceptible to excess water loss at high temperatures. This is because evaporation rate is high through a thin boundary layer and is further enhanced by the open structure of the rosette. Formation of an open rosette structure and small leaves were therefore suggested only to be beneficial under well-watered conditions (Crawford et al. 2012). In water-limited environments, high temperature can result in the closure of stomatal pores. Photosynthesis requires efficient gas exchange between the plants' intercellular spaces and the environment, through stomata. Consequently, mechanisms to prevent water loss under high-temperature stress, such as active stomatal closure, potentially hamper photosynthesis and may raise leaf temperatures even further. Accordingly, *Arabidopsis* mutants which were unable to open or close their stomata had respectively higher and lower leaf temperatures under high-temperature treatment, which correlated with their transpiration rates (Vasseur et al. 2011). Plants facing high-temperature stress therefore have to balance (a) optimizing photosynthesis, (b) avoiding dehydration by excessive water loss, and (c) limiting the risk of overheating. Considering all this, it is perhaps not surprising that water availability strongly interferes with several high-temperature acclimation processes in plants (Mittler 2006; Barnabás et al. 2008).

The effects of drought and high temperature on several traits and their underlying molecular mechanisms are highly interdependent (Rizhsky et al. 2002, 2004; Vile et al. 2012). For instance, heat shock (23–44°C) of tobacco (*Nicotiana tabacum*) and high-temperature transfer of *Arabidopsis* (22–38°C) in well-watered conditions did not directly affect net photosynthesis rates, but evaporation and stomatal conductance were enhanced (Rizhsky et al. 2002, 2004). Under combined heat and drought stress, however, photosynthesis was suppressed and stomata were closed. As expected, this resulted in a rise in leaf temperature. Moreover, increase in specific leaf area (SLA, leaf area per gram dry weight) under heat treatment was abolished, and leaf-to-blade length ratios were more similar to plants under control conditions when heat treatment was combined with drought stress (Vile et al. 2012). Yet, some other architectural traits, such as leaf number and leaf angle, were not altered under additional drought treatment compared to heat stress alone.

Arabidopsis plants grown at 28°C have been reported to develop fewer stomata than 22°C-grown controls, despite showing increased transpiration. No high-temperature-mediated effects on stomatal size or opening were noted (Crawford et al. 2012; Vile et al. 2012). The reason behind these counterintuitive observations is not yet fully understood, but the authors suggest that increased space between stomata may be beneficial for efficient vapor diffusion (Crawford et al. 2012). In conclusion, how plants acclimate to high temperature by adjusting their physiology and morphology is highly dependent on the species under study, the applied temperature regime, and the availability of sufficient water. Acclimation is the sum of effects of several biochemical, physiological, architectural, and morphological traits, which together contribute to an optimal balance between photosynthesis, leaf temperature, and water content.

3.11 Carbohydrate status control of plant acclimation to high temperature

Sugars are not only the primary energy metabolites in plants; they also have important signaling functions in the molecular networks driving responses to environmental change (Rolland et al. 2006; Smeekens et al. 2010). In *Arabidopsis*, for example, sucrose promotes petiole elongation in darkness (Kozuka et al. 2005). Net carbon gain can be affected by changes in temperature through effects on photosynthesis and respiration. Consequently, levels of different sugars can change following high-temperature treatments (Lafta and Lorenzen 1995; Rizhsky et al. 2002, 2004; Kaplan and Guy 2004; Mangelsen et al. 2011). In *Arabidopsis*, end-of-day starch levels were shown to slightly increase with increasing growth temperatures in the range between 12°C and 24°C, while other tested sugars (sucrose, glucose, and fructose) remained unchanged under these conditions (Pyl et al. 2012). Starch levels were always just above depletion at the end of the night, even under high-temperature regimes where sugar demand exceeded availability (Pyl et al. 2012). It was therefore concluded that starch degradation at night is buffered against high temperatures. This homeostatic mechanism allows maximization of growth as all carbon is immediately invested while circumventing the risks of premature carbon starvation during the night.

Carbon gain through photosynthesis is optimized by leaf (re)positioning, relative to the sun. Hyponastic growth was strongly enhanced when high-temperature treatment was combined with low light intensities (van Zanten et al. 2009). Conversely, heat-induced hyponastic growth could be dampened by raising the light levels (Vasseur et al. 2011). These observations inspired Vasseur et al. (2011) to test if acclimation to high temperature is under the control of carbohydrate status. Using the starch-deficient *phosphoglucomutase-1* (*pgm1*) mutant, it was found that light counteracts the deleterious effects of high temperature, by allowing maintenance of high-carbohydrate levels. Indeed, the effect of high temperature on carbon assimilation rate, photosynthesis efficiency, hyponastic growth, and overall carbon nutrient status could all be dampened by growing the plants under higher light intensities (Vasseur et al. 2011).

In summary, carbohydrate status appears to be an important regulator of plant acclimation to high temperature. In situations where high temperature hampers sugar availability, the reduction in carbohydrate levels induces acclimation responses, including hyponastic growth, leading to improvement of the carbohydrate status.

3.12 Thermoperiodic effects on plant growth and architecture

In natural stands, the presence of daylight (photoperiod) and temperature fluctuate diurnally and, in most climates, seasonally. Typically, the photoperiod (day) coincides with warm temperatures relative to the dark period (night). This natural situation is often referred to as positive day/night difference (+DIF) and is favorable for elongation growth in many species (Myster and Moe 1995). Collectively, responses to diurnally fluctuating temperatures are referred to as thermoperiodism (Went 1944).

In *Arabidopsis*, biomass accumulation, RGR, starch accumulation, protein content, and SLA primarily depend on daytime temperatures. Changes in the night temperature hardly affected these traits. When the temperature of the photoperiod is (artificially held) colder relative to the dark period (called negative day/night difference (−DIF)), plant growth is often strongly inhibited (Figure 3.2G, H). −DIF has been shown to effectively reduce internode elongation of various species (Maas and van Hattum 1998; Carvalho et al. 2002; Thingnaes et al. 2003). In horticulture, where plant morphology can directly relate to crop quality, −DIF is frequently applied to enhance crop compactness and is a sustainable alternative for growth-retarding chemicals (Myster and Moe 1995). −DIF also has an inhibitory effect on inflorescence and leaf elongation in *Arabidopsis* (Figure 3.2G, H). Cell elongation and cell numbers have been shown to be reduced with decreasing day temperatures (Thingnaes et al. 2003). Accordingly, a 10°C −DIF resulted in a 20% decrease of final leaf length compared to + DIF. Despite the economic importance, full mechanistic understanding of the growth reduction under −DIF is still lacking and seems species specific. Increased gibberellin catabolism under −DIF has been described in pea (*Pisum sativum*) (Stavang et al. 2005) but could not be detected in begonia (*Begonia x hiemalis*) (Myster et al. 1997) or *Arabidopsis* (Thingnaes et al. 2003). The −DIF growth response in *Arabidopsis* is, however, associated with reduced auxin levels (Thingnaes et al. 2003). As with several other responses described in this chapter, photoreceptors are involved in growth inhibition under −DIF. Red light has been shown to enhance the −DIF response in multiple species, while supplemental far-red light reduced this effect (Moe et al. 1991; Erwin and Heins 1995). In accordance with these observations, *phyB* mutants in *Arabidopsis* (Thingnaes et al. 2008) and cucumber (*Cucumis sativus*) (Patil and Moe 2009) are insensitive to −DIF treatment.

Alternating temperatures provide a useful tool to study the interaction between light and temperature in the control of rhythmic plant growth and development. Using −DIF, it was shown that certain diurnal processes are preferentially regulated by temperature over light. For example, diurnal ethylene emissions in sorghum (*Sorghum bicolor*) always peaked during the subjective warm period regardless of the photoperiod (Finlayson et al. 1998). In addition, both light and temperature entrain the circadian clock (Jones 2009; Chapter 6). Diurnal leaf movements at different temperatures are affected through temperature-compensation mechanisms of the clock (Edwards et al. 2005). The expression patterns of different clock-controlled genes display unequal phase shifts and became desynchronized under continuous (free-running) conditions following entrainment to −DIF (Michael et al. 2003). Disrupted clock function may therefore underlie the reduced growth phenotypes observed in these conditions.

3.13 High-temperature effects on the floral transition

The switch from vegetative growth to flowering is one of the major developmental transitions in the plant life cycle. In many species the timing of flowering is influenced by environmental factors which serve to communicate the time of year and growth conditions favorable for sexual reproduction. Temperature, together with photoperiod, is among the most important. In many perennial and biennial/winter-annual species, flowering is accelerated through prolonged cold exposure in a process called vernalization (Michaels and Amasino 2000). This is discussed in Chapter 4. Moderate changes in ambient growth temperature can also have dramatic effects on flowering time. In general, increasing temperature promotes flowering with a shorter vegetative phase and by increasing RGR (Figure 3.2A, B). In contrast, lowering temperature significantly delays flowering. Several studies on mean flowering times and/or first flowering dates of wild plant species in the UK over several decades show that these phenological events are significantly related to ambient temperature. It is expected that flowering of the species examined will advance by 5–25 days under the current climate change models, which predict an increase in average temperature of +2.5°C (Fitter et al. 1995; Sparks and Carey 1995; Sparks et al. 2000; Fitter and Fitter 2002). As a result, 150 to 200 species in the UK may now be flowering on average 15 days earlier than in the very recent past (Fitter and Fitter 2002). Some evidence suggests that the temperature dependency of flowering time is diurnally regulated. In *Arabidopsis*, high temperature at night is a more potent floral inducer than high temperature during the day (Thingnaes et al. 2003). In other plants, such as orchids, this relationship is reversed, and high daytime temperature is a more effective floral-promoting signal (Blanchard and Runkle 2006).

Rising average global temperatures are predicted to significantly hamper agricultural productivity worldwide (Battisti and Naylor 2009), in part through temperature effects on the flowering process. For example, temperate cereals, including the staple food crops wheat and barley, progress more rapidly through early stages of reproductive development when grown at higher temperatures under long-day (LD) conditions (Hemming et al. 2012). As a result, grain number is reduced (Fischer 1985; Rawson and Richards 1993). A rise in average temperature can also affect the timing of floral transition, further reducing yield by increasing the risk of frost- or heat-related damage to developing grains. To maintain crop yields in a changing climate, it is therefore of uttermost importance to understand how reproductive development responds to high temperature.

Work in *Arabidopsis* has defined a complex network of regulatory pathways that controls flowering time in response to diverse environmental signals. Ultimately, developmental and environmental signals converge at the shoot apex and quantitatively regulate an overlapping set of common targets, the floral pathway integrators (FPI), whose activities evoke a florally determined state in the shoot apical meristem (SAM). In *Arabidopsis*, FPI genes include *FLOWERING LOCUS T* (*FT*) and *SUPPRESSOR OF OVEREXPRESSION OF CONSTANS1* (*SOC1*). Their combined effects are conveyed to floral meristem identity (FMI) genes at the SAM that trigger the flowering process, such as *LEAFY (LFY)* and *APETALA1* (*AP1*) (reviewed by Fornara et al. 2010). Floral induction is antagonized by members of the floral repression pathway, which is considered to regulate the competence of the SAM to floral-promoting signals. In *Arabidopsis*, these members include the MADS-box transcription factors FLOWERING LOCUS C (FLC), FLOWERING LOCUS M (FLM), and SHORT VEGETATIVE PHASE (SVP). FLC, which integrates prolonged cold signals (vernalization) and endogenous inputs from the so-called autonomous pathway, FLM, and SVP delay flowering by repressing transcription of FPI genes (reviewed by Yant et al. 2009; Jarillo and Piñeiro 2011).

To elucidate the molecular mechanisms underlying ambient temperature effects on flowering, genetic and natural variation screens have been performed (Blázquez et al. 2003; Lempe et al. 2005; Balasubramanian et al. 2006; Lee et al. 2007). These have revealed a thermosensory flowering pathway that coordinates the response to daily growth temperatures. Over the past few years, several studies identified additional components that play a major role in temperature perception and signaling in flowering. These include several well-characterized flowering time regulators, such as FLM; SVP; the autonomous pathway components FCA, FVE, and FY; EARLY FLOWERING 3 (ELF3); TERMINAL FLOWER 1 (TFL1); several ambient temperature-responsive microRNAs (miRNAs); as well as PIF4 and the histone H2A variant H2A.Z (Blázquez et al. 2003; Balasubramanian and Weigel 2006; Balasubramanian et al. 2006; Lee et al. 2007, 2010; Strasser et al. 2009; Kumar and Wigge 2010; Kim et al. 2011; Kumar et al. 2012). From these studies it has also become clear that temperature-mediated floral timing involves not just a single but several different thermosensory pathways that act mostly in parallel and converge at the level of *FT* and, most likely, additional downstream targets.

In *Arabidopsis*, a facultative LD species, high temperature potently triggers flowering in both inductive and noninductive photoperiods (Figure 3.2A, B). In noninductive short-day (SD) photoperiods, PIF4 is essential for the thermal induction of flowering. Mutants deficient in PIF4 fail to show temperature-mediated acceleration of flowering under these conditions (Kumar et al. 2012). PIF4 directly activates *FT* by binding the *FT* promoter in a temperature-dependent manner. The temperature-mediated activation of *FT* by PIF4 is controlled at the level of chromatin accessibility. As described previously, at low temperatures H2A.Z nucleosomes have a high level of occupancy. This includes the PIF4-binding site in the *FT* promoter. At higher temperatures, H2A.Z nucleosome occupancy declines, suggesting that the presence of H2A.Z nucleosomes is limiting for PIF4 binding (Kumar et al. 2012).

FLM has also been found to modulate the sensitivity to temperature (Balasubramanian et al. 2006). FLM is the major repressor that suppresses flowering in SD. Whereas PIF4 acts as an activator of *FT*, FLM functions as a repressor. Higher ambient temperature results in changes in the splicing pattern of *FLM* which likely leads to reduction in the splice form that represses *FT*. This subsequently leads to increased *FT* expression and ultimately results in flowering (Balasubramanian and Weigel 2006).

In floral-inductive conditions, PIF4 and FLM do not appear to play a significant role in thermosensory regulation of flowering. Instead, several other genes have been identified that affect temperature sensitivity in LD and/or continuous light (CL) conditions. Whether and how these are functionally interrelated is still largely unknown. For at least two of the identified genes, *TFL1* and *ELF3*, it is clear that they function in two genetically distinguishable pathways. Both act as suppressors of flowering at lower temperatures. *ELF3* interacts with the photoperiod pathway and circadian clock components in the leaves, whereas *TFL1* regulates the FPI *SOC1* in the shoot apex (Strasser et al. 2009). Autonomous pathway mutants such as *fca*, *fve*, *fpa*, *flk*, and *ld* are insensitive to the promotion of flowering by increasing temperature, suggesting that the corresponding genes may have a function in this response (Blázquez et al. 2003; Lee et al. 2007). The failure of these mutants to respond to thermal induction appears to be largely due to their elevated *FLC* levels, (Blázquez et al. 2003; Balasubramanian et al. 2006; Lee et al. 2007). The temperature-dependent functions of FCA and FVE are mediated by SVP (Lee et al. 2007). To exert an effect on flowering, SVP negatively regulates the expression of both *FT* and *SOC1* by directly binding their genomic loci. In addition, SVP has been shown to act as a direct transcriptional inhibitor of miR172 (Cho et al. 2012). This miRNA is a potent floral activator and its accumulation is ambient temperature responsive. Plants overexpressing miR172 show a temperature-independent early-flowering phenotype, suggesting that modulation of miR172 expression leads to temperature insensitivity (Lee et al. 2010). Intriguingly, processing of primary miR172 transcript (pri-miR172) is controlled in an ambient temperature-dependent manner by FCA (Jung et al. 2012). The study that identified miR172 to be an ambient temperature-regulated miRNA also identified miR156 (Lee et al. 2010). Expression of miR156 is enhanced by lower temperatures. miR156 is known to target members of the *SQUAMOSA PROMOTER-BINDING PROTEIN-LIKE* (*SPL*) gene family, including *SPL3*. SPL3 functions as a direct upstream activator of FT. Changes in ambient temperature therefore cause alterations in the expression of miR156, which negatively regulates *SPL3* and *FT* and the floral transition process (Kim et al. 2012). Although miR172 and miR156 play distinct roles in floral transition, signaling cross talk between the two miRNA pathways has been reported, mediated by the SPL3/4/5 genes (Jung et al. 2011). It is therefore currently not clear whether both miRNAs act via single or parallel pathways in the regulation of ambient temperature-responsive flowering. Nevertheless, it is clear that the altered accumulation of specific miRNAs in reaction to changing ambient temperatures is important in the generation of a thermal response. Since several other ambient temperature-responsive miRNAs have been identified (Lee et al. 2010), miRNA accumulation might provide a more general signaling mechanism linking temperature sensing to developmental decisions in the plant life cycle.

Acknowledgments

We thank Nicole Rodenburg and Mark D. van Hoogdalem (Utrecht University) for photos. Our work on heat acclimation is supported by VENI grant 863.11.008 of the Netherlands Organization for Scientific Research to MvZ, our work on −DIF is funded by the Top Technological Institute Green Genetics (TTI-GG) (Project: 2CFL009RP) (RB), and our work on temperature-dependent flowering by TTI-GG Project: 2CC034RP (MCGP).

References

Abreu ME, Munné-Bosch S (2008) Hyponastic leaf growth decreases the photoprotective demand, prevents damage to photosystem II and delays leaf senescence in *Salvia broussonetii* plants. Physiol. Plant 134: 369–379.

Armstrong AF, Logan DC, Tobin AK, O'Toole P, Atkin OK (2006a) Heterogeneity of plant mitochondrial responses underpinning respiratory acclimation to the cold in *Arabidopsis thaliana* leaves. Plant Cell Environ. 29: 940–949.

Armstrong AF, Logan DC, Atkin OK (2006b) On the developmental dependence of leaf respiration: Responses to short- and long-term changes in growth temperature. Am. J. Bot. 93: 1633–1639.

Armstrong AF, Wardlaw KD, Atkin OK (2007) Assessing the relationship between respiratory acclimation to the cold and photosystem II redox poise in *Arabidopsis thaliana*. Plant Cell Environ. 30: 1513–1522.

Athanasiou K, Dyson BC, Webster RE, Johnson GN (2010) Dynamic acclimation of photosynthesis increases plant fitness in changing environments. Plant Physiol. 152: 366–373.

Atkin OK, Tjoelker MG (2003) Thermal acclimation and the dynamic response of plant respiration to temperature. Trends Plant Sci. 8: 343–351.

Atkin OK, Scheurwater I, Pons TL (2006a) High thermal acclimation potential of both photosynthesis and respiration in two lowland *Plantago* species in contrast to an alpine congeneric. Glob. Change Biol. 12: 500–515.

Atkin OK, Loveys BR, Atkinson LJ, Pons TL (2006b) Phenotypic plasticity and growth temperature: Understanding interspecific variability. J. Exp. Bot. 57: 267–281.

Atkin OK, Scheurwater I, Pons TL (2007) Respiration as a percentage of daily photosynthesis in whole plants is homeostatic at moderate, but not high, growth temperatures. New Phytol. 174: 367–380.

Balasubramanian S, Weigel D (2006) Temperature induced flowering in *Arabidopsis thaliana*. Plant Signal. Behav. 1: 227–228.

Balasubramanian S, Sureshkumar S, Lempe J, Weigel D (2006) Potent induction of *Arabidopsis thaliana* flowering by elevated growth temperature. PLoS Genet. 2: 980–989.

Barnabás B, Jäger K, Fehér A (2008) The effect of drought and heat stress on reproductive processes in cereals. Plant Cell Environ. 31: 11–38.

Battisti DS, Naylor RS (2009) Historical warnings of future food insecurity with unprecedented seasonal heat. Science 323: 240–244.

Benschop JJ, Millenaar FF, Smeets ME, van Zanten M, Voesenek LACJ, Peeters AJM (2007) ABA antagonizes ethylene-induced hyponastic growth in *Arabidopsis*. Plant Physiol. 143: 1013–1023.

Berry J, Björkman O (1980) Photosynthetic response and adaptation to temperature in higher plants. Ann. Rev. Plant Physiol. 31: 491–543.

Blanchard MG, Runkle ES (2006) Temperature during the day, but not during the night, controls flowering of *Phalaenopsis* orchids. J. Exp. Bot. 57: 4043–4039.

Blázquez MA, Ahn JH, Weigel DA (2003) Thermosensory pathway controlling flowering time in *Arabidopsis thaliana*. Nat. Genet. 33: 168–171.

Carberry PS, Albrecht DG (1990) Germination and elongation of the hypocotyl and radicle of kenaf (*Hibiscus cannabinus*) in response to temperature. Field Crop. Res. 24: 227–240.

Carvalho SMP, Heuvelink E, Cascais R, Van Kooten O (2002) Effect of day and night temperature on internode and stem length in chrysanthemum: Is everything explained by DIF? Ann. Bot. (Lond). 90: 111–118.

Cho HJ, Kim JJ, Lee JH, Kim W, Jung JH, Park CM, Ahn JH (2012) SHORT VEGETATIVE PHASE (SVP) protein negatively regulates miR172 transcription via direct binding to the pri-miR172a promoter in *Arabidopsis*. FEBS Lett. 586: 2332–2337.

Clarke SM, Cristescu SM, Miersch O, Harren FJM, Wasternack C, Mur LAJ (2009) Jasmonates act with salicylic acid to confer basal thermotolerance in *Arabidopsis thaliana*. New Phytol. 182: 175–87.

Crafts-Brandner SJ, Salvucci ME (2000) Rubisco activase constrains the photosynthetic potential of leaves at high temperature and CO_2. Proc. Natl. Acad. Sci. USA 97: 13430–13435.

Crawford AJ, McLachlan DH, Hetherington AM, Franklin KA (2012) High temperature exposure increases plant cooling capacity. Curr. Biol. 22: 396–397.

Edwards KD, Lynn JR, Gyula P, Nagy F, Millar AJ (2005). Natural allelic variation in the temperature-compensation mechanisms of the *Arabidopsis thaliana* circadian clock. Genetics 170: 387–400.

Erwin JE, Heins RD (1995) Thermomorphogenic responses in stem and leaf development. Hortscience 30: 940–949.

Falster DS, Westoby M (2003) Leaf size and angle vary widely across species: What consequences for light interception? New Phytol. 158: 509–525.

Finlayson SA, Lee IJ, Morgan PW (1998) Phytochrome B and the regulation of circadian ethylene production in sorghum. Plant Physiol. 116: 17–25.

Fischer RA (1985) Number of kernels in wheat crops and the influence of solar radiation and temperature. J. Agric. Sci. 105: 447–461.

Fitter AH, Fitter RS (2002) Rapid changes in flowering time in British plants. Science 296: 1689–1691.

Fitter AH, Fitter RSR, Harris ITB, Williamson MH (1995) Relationships between first flowering date and temperature in the flora of a locality in central England. Funct. Ecol. 9: 55–60.

Flood PJ, Harbinson J, Aarts MG (2011) Natural genetic variation in plant photosynthesis. Trends Plant Sci. 16: 327–335.

Foreman J, Johansson H, Hornitschek P, Josse EM, Fankhauser C, Halliday KJ (2011) Light receptor action is critical for maintaining plant biomass at warm ambient temperatures. Plant J. 65: 441–452.

Fornara F, de Montaigu A, Coupland G (2010) SnapShot: Control of flowering in *Arabidopsis*. Cell 141: 550.e1–550.e2.

Fournier-Level A, Korte A, Cooper MD, Nordborg M, Schmitt J, Wilczek AM (2011) A map of local adaptation in *Arabidopsis thaliana*. Science 334: 86–89.

Franklin KA (2009) Light and temperature signal crosstalk in plant development. Curr. Opin. Plant Biol. 12: 63–68.

Franklin KA, Lee SH, Patel D, Kumar SV, Spartz AK, Gu C, Ye S, Yu P, Breen G, Cohen JD, Wigge PA, Gray WM (2011) Phytochrome-interacting factor 4 (PIF4) regulates auxin biosynthesis at high temperature. Proc. Natl. Acad. Sci. USA 108: 20231–20235.

Fu QA, Ehleringer JR (1989) Heliotropic leaf movements in common beans controlled by air temperature. Plant Physiol. 91: 1162–1167.

Gifford RM (1995) Whole plant respiration and photosynthesis of wheat under increased CO_2 concentration and temperature: Long-term vs short-term distinctions for modelling. Glob. Change Biol. 1: 385–396.

Gorsuch PA, Pandey S, Atkin OK (2010) Temporal heterogeneity of cold acclimation phenotypes in *Arabidopsis* leaves. Plant Cell Environ. 33: 244–258.

Granier C, Massonnet C, Turc O, Muller B, Chenu K, Tardieu F (2002) Individual leaf development in *Arabidopsis thaliana*: a stable thermal-time-based programme. Ann. Bot (Lond). 89: 595–604.

Gray WM, Ostin A, Sandberg G, Romano CP, Estelle M (1998). High temperature promotes auxin-mediated hypocotyl elongation in *Arabidopsis*. Proc. Natl. Acad. Sci. USA 95: 7197–7202.

Hancock AM, Brachi B, Faure N, Horton MW, Jarymowycz LB, Sperone FG, Toomajian C, Roux F, Bergelson J (2011) Adaptation to climate across the *Arabidopsis thaliana* genome. Science 334: 83–86.

Hangarter RP (1997) Gravity, light and plant form. Plant Cell Environ. 20: 796–800.

Hemming MN, Walford SA, Fieg S, Dennis ES, Trevaskis B (2012) Identification of high-temperature-responsive genes in cereals. Plant Physiol. 158: 1439–1450.

Hikosaka K (1997) Modelling optimal temperature acclimation of the photosynthetic apparatus in C-3 plants with respect to nitrogen use. Ann. Bot. (Lond.) 80: 721–730.

Hoffmann MH (2002) Biogeography of *Arabidopsis thaliana* (L.) Heynh. (Brassicaceae). J. Biogeography. 29: 125–134.

Hong SW, Vierling E (2000) Mutants of *Arabidopsis thaliana* defective in the acquisition of tolerance to high temperature stress. Proc. Natl. Acad. Sci. USA 97: 4392–4397.

Hopkins R, Schmitt J, Stinchcombe JR (2008) A latitudinal cline and response to vernalization in leaf angle and morphology in *Arabidopsis thaliana* (Brassicaceae). New Phytol. 179: 155–164.

Huq E, Quail PH (2002) PIF4, a phytochrome-interacting bHLH factor, functions as a negative regulator of phytochrome B signaling in *Arabidopsis*. EMBO J. 22: 2441–2550.

Jarillo JA, Piñeiro M (2011) Timing is everything in plant development. The central role of floral repressors. Plant Sci. 181: 364–78.

Jones M (2009) Entrainment of the *Arabidopsis* circadian clock. J. Plant Biol. 52: 202–209.

Jung JH, Seo PJ, Kang SK, Park CM (2011) miR172 signals are incorporated into the miR156 signaling pathway at the SPL3/4/5 genes in *Arabidopsis* developmental transitions. Plant Mol. Biol. 76: 35–45.

Jung JH, Seo PJ, Ahn JH, Park CM (2012) *Arabidopsis* RNA-binding protein FCA regulates microRNA172 processing in thermosensory flowering. J. Biol. Chem. 287: 16007–16016.

Kaplan F, Guy CL (2004) Beta-amylase induction and the protective role of maltose during temperature shock. Plant Physiol. 135: 1674–1684.

Kesari R, Lasky JR, Villamor JG, Des Marais DL, Chen Y-JC, Liu T-W, Lin W, Juenger TE, Verslues PE (2012) Intron-mediated alternative splicing of *Arabidopsis* P5CS1 and its association with natural variation in proline and climate adaptation. Proc. Natl. Acad. Sci. USA 103: 9197–9202.

Kim W, Ahn HJ, Chiou TJ, Ahn JH (2011) The role of the miR399-PHO_2 module in the regulation of flowering time in response to different ambient temperatures in *Arabidopsis thaliana*. Mol. Cells 32: 83–88.

Kim JJ, Lee JH, Kim W, Jung HS, Huijser P, Ahn JH (2012) The microRNA156-SQUAMOSA PROMOTER BINDING PROTEIN-LIKE3 module regulates ambient temperature-responsive flowering via FLOWERING LOCUS T in *Arabidopsis*. Plant Physiol. 159: 461–78.

King DA (1997) The functional significance of leaf angle in Eucalyptus. Aust. J. Bot. 45: 619–639.

Koini MA, Alvey L, Allen T, Tilley CA, Harberd NP, Whitelam GC, Franklin KA (2009) High temperature-mediated adaptations in plant architecture require the bHLH transcription factor PIF4. Curr. Biol. 19: 408–413.

Kotak S, Larkindale J, Lee U, von Koskull-Döring P, Vierling E, Scharf KD (2007) Complexity of the heat stress response in plants. Curr. Opin. Plant Biol. 10: 310–316.

Kozuka T, Horiguchi G, Kim GT, Ohgishi M, Sakai T, Tsukaya H (2005) The different growth responses of the *Arabidopsis thaliana* leaf blade and the petiole during shade avoidance are regulated by photoreceptors and sugar. Plant Cell Physiol. 46: 213–223.

Kumar SV, Wigge PA (2010). H2A.Z-containing nucleosomes mediate the thermosensory response in *Arabidopsis*. Cell 140: 136–147.

Kumar SV, Lucyshyn D, Jaeger KE, Alós E, Alvey E, Harberd NP, Wigge PA (2012) Transcription factor PIF4 controls the thermosensory activation of flowering. Nature 21: 484: 242–245.

Kurek I, Chang TK, Bertain SM, Madrigal A, Liu L, Lassner MW, Zhu GH (2007) Enhanced thermostability of *Arabidopsis* Rubisco activase improves photosynthesis and growth rates under moderate heat stress. Plant Cell 19: 3230–3241.

Lafta AM, Lorenzen H (1995) Effect of high temperature on plant growth and carbohydrate metabolism in potato. Plant Physiol. 109: 637–643.

Lambers H, Chapin III FS, Pons TL (2008) Plant physiological ecology. 2nd edn. Springer, New York.

Larkindale J, Knight MR (2002) Protection against heat stress-induced oxidative damage in *Arabidopsis* involves calcium, abscisic acid, ethylene, and salicylic acid. Plant Physiol. 128: 682–695.

Larkindale J, Hall JD, Knight MR, Vierling E (2005) Heat stress phenotypes of *Arabidopsis* mutants implicate multiple signaling pathways in the acquisition of thermotolerance. Plant Physiol. 138: 882–897.

Lee JH, Yoo SJ, Park SH, Hwang I, Lee JS, Ahn JH (2007) Role of SVP in the control of flowering time by ambient temperature in *Arabidopsis*. Genes Dev. 21: 397–402.

Lee H, Yoo SJ, Lee JH, Kim W, Yoo SK, Fitzgerald H, Carrington JC, Ahn JH (2010) Genetic framework for flowering-time regulation by ambient temperature-responsive miRNAs in *Arabidopsis*. Nucleic Acids Res. 38: 3081–3093.

Lempe J, Balasubramanian S, Sureshkumar S, Singh A, Schmid M, Weigel D (2005) Diversity of flowering responses in wild *Arabidopsis thaliana* strains. PLoS Genet. 1:109–118.

Lorrain S, Allen T, Duek PD, Whitelam GC, Fankhauser C (2008) Phytochrome-mediated inhibition of shade avoidance involves degradation of growth-promoting bHLH transcription factors. Plant J. 53: 312–323.

Loveys BR, Scheurwater I, Pons TL, Fitter AH, Atkin OK (2002) Growth temperature influences the underlying components of relative growth rate: an investigation using inherently fast- and slow-growing plant species. Plant Cell Environ. 25: 975–988.

Maas FM, van Hattum J (1998) Thermomorphogenic and photomorphogenic control of stem elongation in fuchsia is not mediated by changes in responsiveness to gibberellins. J. Plant Growth Regul. 17: 39–45.

Mangelsen E, Kilian J, Harter K, Jansson C, Wanke D, Sundberg E (2011) Transcriptome analysis of high-temperature stress in developing barley caryopses: Early stress responses and effects on storage compound biosynthesis. Mol. Plant 4: 97–115.

Mazzella MA, Bertero D, Casal JJ (2000) Temperature-dependent internode elongation in vegetative plants of *Arabidopsis thaliana* lacking phytochrome B and cryptochrome 1. Planta 210: 497–501.

McClung R, Davis SJ (2010) Ambient thermometers in plants: From physiological outputs towards mechanisms of thermal sensing. Curr. Biol. 1086–1092.

Michael TP, Salome PA, McClung CR (2003) Two *Arabidopsis* circadian oscillators can be distinguished by differential temperature sensitivity. Proc. Natl. Acad. Sci. USA 100: 6878–6883.

Michaels SD, Amasino RA (2000) Memories of winter: Vernalization and the competence to flower. Plant Cell Environ. 23: 1145–1153.

Millenaar FF, Cox MC, van Berkel YE, Welschen RA, Pierik R, Voesenek LACJ, Peeters AJM (2005) Ethylene-induced differential growth of petioles in *Arabidopsis*. Analyzing natural variation, response kinetics, and regulation. Plant Physiol. 137: 998–1008.

Millenaar FF, van Zanten M, Cox MCH, Pierik R, Voesenek LACJ, Peeters AJM (2009) Differential petiole growth in *Arabidopsis thaliana*: Photocontrol and hormonal regulation. New Phytol. 184: 141–152.

Mittler R (2006) Abiotic stress, the field environment and stress combination. Trends Plant Sci. 11: 15–19.

Mittler R, Finka A, Goloubinoff P (2012) How do plants feel the heat? Trends Biochem. Sci. 37: 118–125.

Moe R, Heins RD, Erwin J (1991) Stem elongation and flowering of the long-day plant *campanula-isophylla moretti* in response to day and night temperature alternations and light quality. Sci. Hortic. 48: 141–151.

Monson RK, Holland EA (2001) Biospheric trace gas fluxes and their control over tropospheric chemistry. Annu. Rev. Ecol. Syst. 32: 547–576.

Monson RK, Grote R, Niinemets U, Schnitzler JP (2012) Modeling the isoprene emission rate from leaves. New Phytol. 195: 541–559.

Mullen JL, Weinig C, Hangarter RP (2006) Shade avoidance and the regulation of leaf inclination in *Arabidopsis*. Plant Cell Environ. 29: 1099–1106.

Myster J, Moe R (1995) Effect of diurnal temperature alternations on plant morphology in some greenhouse crops – a mini review. Sci. Hortic. 62: 205–215.

Myster J, Ernstsen A, Junttila O, Moe R (1997) Thermo- and photoperiodicity and involvement of gibberellins during day and night cycle on elongation growth of *Begonia x hiemalis fotsch*. J. Plant Growth Reg. 16: 189–196.

Nicotra AB, Atkin OK, Bonser SP, Davidson M, Finnegan EJ, Mathesius U, Poot MD, Purugganan AM, Richards CL, Valladares F, Van Kleunen M (2010) Plant phenotypic plasticity in a changing climate. Trends Plant Sci. 15: 684–669.

Orbović V, Poff KL (2007) Effect of temperature on growth and phototropism of *Arabidopsis thaliana* seedlings. J. Plant Growth Regul. 26: 222–228.

Parent B, Tardieu F (2012) Temperature responses of developmental processes have not been affected by breeding in different ecological areas for 17 crop species. New Phytol. 194: 760–774.

Parent B, Turc O, Gibon Y, Stitt M, Tardieu F (2010) Modelling temperature-compensated physiological rates, based on the co-ordination of responses to temperature of developmental processes. J. Exp. Bot. 61: 2057–2069.

Patel D, Franklin KA (2009) Temperature-regulation of plant architecture. Plant Signal. Behav. 4: 577–579.

Patel D, Basu M, Hayes S, Majlath I, Hetherington FM, Tschaplinski T, Franklin KA (2013) Temperature-dependent shade avoidance involves the receptor-like kinase, ERECTA. Plant J. 73: 980–992.

Patil GG, Moe R (2009) Involvement of phytochrome B in DIF mediated growth in cucumber. Sci. Hortic. 122: 164–170.

Pauli H, Gottfried M, Dullinger S, Abdaladze O, Akhalkatsi M, Benito Alonso JL, Coldea G, Dick J, Erschbamer B, Fernández Calzado R, Ghosn D, Holten JI, Kanka R, Kazakis G, Kollár J, Larsson P, Moiseev P, Moiseev D, Molau U, Molero Mesa J, Nagy L, Pelino G, Pu⊠ca⊠ M, Rossi G, Stanisci A, Syverhuset AO, Theurillat JP, Tomaselli M, Unterluggauer P, Villar L, Vittoz P, Grabherr G (2012) Recent plant diversity changes on Europe's mountain summits. Science 20: 353–355.

Pecinka A, Dinh HQ, Baubec T, Rosa M, Lettner N, Mittelsten-Scheid O (2010) Epigenetic regulation of repetitive elements is attenuated by prolonged heat stress in *Arabidopsis*. Plant Cell 22: 3118–3129.

Peña-Castro J, van Zanten M, Lee SC, Patel M, Voesenek LACJ, Fukao T, Bailey-Serres J (2011) Expression of rice SUB1A and SUB1C transcription factors in *Arabidopsis* uncovers flowering inhibition as a submergence-tolerance mechanism. Plant J. 67: 434–446.

Penfield S (2008) Temperature perception and signal transduction in plants. New Phytol. 179: 615–628.

Pierik R, Millenaar FF, Peeters AJM, Voesenek LACJ (2005) New perspectives in flooding research: The use of shade avoidance and *Arabidopsis thaliana*. Ann. Bot. (Lond). 96: 533–540.

Polko JK, van Zanten M, van Rooij JA, Marée AF, Voesenek LACJ, Peeters AJM, Pierik R (2012) Ethylene-induced differential petiole growth in *Arabidopsis thaliana* involves local microtubule reorientation and cell expansion. New Phytol. 193: 339–348.

Pons TL (2012) Interaction of temperature and irradiance effects on photosynthetic acclimation in two accessions of *Arabidopsis thaliana*. Photosynth. Res. 113: 207–219.

Poorter H, Niinemets U, Poorter L, Wright IJ, Villar R (2009) Causes and consequences of variation in leaf mass per area (LMA): A meta-analysis. New Phytol. 182: 565–588.

Poorter H, Niklas KJ, Reich PB, Oleksyn J, Poot P, Mommer L (2012) Biomass allocation to leaves, stems and roots: Meta-analyses of interspecific variation and environmental control. New Phytol. 193: 30–50.

Probert RJ (2000) The role of temperature in the regulation of seed dormancy and germination. In: Fenner M (ed.). Seeds: The ecology of regeneration in plant communities. CAB International, Wallingford, pp. 261–292.

Pyl E-T, Piques M, Ivakov A, Schulze W, Ishihara H, Stitt M, Sulpice R (2012) Metabolism and growth in *Arabidopsis* depend on the daytime temperature but are temperature-compensated against cool nights. Plant Cell 24: 2443–2469.

Rawson HM, Richards RA (1993) Effects of high temperature and photoperiod on floral development in wheat isolines differing in vernalisation and photoperiod genes. Field Crops Res. 32: 181–192.

Richards CL, Rosas U, Banta J, Bhambhra N, Purugganan MD (2012). Genome-wide patterns of *Arabidopsis* gene expression in nature. PLoS Genet. 8: e1002662.

Rizhsky L, Hongjian L, Mittler R (2002) The combined effect of drought stress and heat shock on gene expression in tobacco. Plant Physiol. 130: 1143–1151.

Rizhsky L, Liang H, Shuman J, Shulaev V, Davletova S, Mittler R (2004) When defense pathways collide. The response of *Arabidopsis* to a combination of drought and heat stress. Plant Physiol. 134: 1683–1696.

Rolland F, Baena-Gonzalez E, Sheen J (2006) Sugar sensing and signalling in plants: Conserved and novel mechanisms. Annu. Rev. Plant Biol. 57: 675–709.

Sage RF, Kubien DS (2007) The temperature response of C(3) and C(4) photosynthesis. Plant Cell Environ. 30: 1086–1106.

Schmuths H, Bachmann K, Weber WE, Horres R, Hoffmann MH (2006) Effects of preconditioning and temperature during germination of 73 natural accessions of *Arabidopsis thaliana*. Ann Bot. (Lond.) 97: 623–634.

Sexton JP, McIntyre PJ, Angert AL, Rice KJ (2009) Evolution and ecology and 586 species range limits. Annu. Rev. Ecol. Syst. 40: 415–436.

Seymour RS (2001) Biophysics and physiology of temperature regulation in thermogenic flowers. Biosci. Rep. 21: 223–236.

Sharkey TD, Zhang R (2010) High temperature effects on electron and proton circuits of photosynthesis. J. Integr. Plant Biol. 52: 712–722.

Smallwood M, Bowles DJ (2002) Plants in a cold climate. Philos. Trans. R. Soc. Lond. B 357: 831–847.

Smeekens S, Ma J, Hanson J, Rolland F (2010) Sugar signals and molecular networks controlling plant growth. Curr. Opin. Plant Biol. 13: 274–249.

Smith AM, Stitt M (2007) Coordination of carbon supply and plant growth. Plant Cell Environ. 30: 1126–1149.

Sparks TH, Carey PD (1995) The responses of species to climate over two centuries: an analysis of the Marsham phenological record, 1736–1947. J. Ecol. 83: 321–329.

Sparks TH, Jeffree EP, Jeffree CE (2000) An examination of the relationship between flowering times and temperature at the national scale using long-term phenological records from the UK. Int. J. Biometeorol. 44: 82–87.

Stavang JA, Lindgard B, Erntsen A, Lid SE, Moe R, Olsen JE (2005) Thermoperiodic stem elongation involves transcriptional regulation of gibberellin deactivation in pea. Plant Physiol. 138: 2344–2353.

Stavang JA, Gallego-Bartolomé J, Gómez MD, Yoshida S, Asami T, Olsen JE, García-Martínez JL, Alabadí D, Blázquez MA (2009) Hormonal regulation of temperature-induced growth in *Arabidopsis*. Plant J. 60: 589–601.

Stitt M, Hurry V (2002) A plant for all seasons: Alterations in photosynthetic carbon metabolism during cold acclimation in *Arabidopsis*. Curr. Opin. Plant Biol. 5: 199–206.

Stoddart JL, Tapster SM, Jones TWA (1978) Temperature dependence of the gibberellin response in lettuce hypocotyls. Planta 141: 283–288.

Strasser B, Alvarez MJ, Califano A, Cerdán PD (2009) A complementary role for ELF3 and TFL1 in the regulation of flowering time by ambient temperature. Plant J. 58: 629–640.

Sun J, Qi L, Li Y, Chu J, Li C (2012) PIF4-Mediated activation of YUCCA8 expression integrates temperature into the auxin pathway in regulating *Arabidopsis* hypocotyl growth. PLoS Genet. 8: e1002594.

Terashima I, Hanba YT, Tholen D, Niinemets Ü (2011) Leaf functional anatomy in relation to photosynthesis. Plant Physiol. 155: 108–116.

Thingnaes E, Torre S, Ernstsen A, Moe R (2003) Day and night temperature responses in *Arabidopsis*: Effects on gibberellin and auxin content, cell size, morphology and flowering time. Ann. Bot. (Lond). 92: 601–612.

Thingnaes E, Torre S, Moe R (2008) The role of phytochrome B, D and E in thermoperiodic responses of *Arabidopsis thaliana*. Plant Growth Reg. 56: 53–59.

Thuiller W, Lavorel S, Araújo MB, Sykes MT, Prentice IC (2007) Climate change threats to plant diversity in Europe. Proc. Natl. Acad. Sci. USA 102: 8245–8250.

Tittel-Elmer M, Bucher E, Broger L, Mathieu O, Paszkowski J, Vaillant I (2010) Stress-induced activation of heterochromatic transcription. PLoS Genet. 6: e1001175.

van Zanten M, Voesenek LACJ, Peeters AJM, Millenaar FF (2009) Hormone- and light-mediated regulation of heat-induced differential petiole growth in *Arabidopsis thaliana*. Plant Physiol. 151: 1446–1458.

van Zanten M, Pons TL, Janssen JAM, Voesenek LACJ, Peeters AJM (2010) On the relevance and control of leaf angle. Crit. Rev. Plant Sci. 29: 300–316.

van Zanten M, Ritsema T, Polko JK, Leon-Reyes A, Voesenek LACJ, Millenaar FF, Pieterse CMJ, Peeters AJM (2012) Modulation of ethylene and heat-controlled hyponastic leaf movement in *Arabidopsis thaliana* by the plant defense hormones jasmonate and salicylate. Planta 235: 677–685.

Vasseur F, Pantin F, Vile D (2011) Changes in light intensity reveal a major role for carbon balance in *Arabidopsis* in response to high temperature. Plant Cell Environ. 34: 1563–1576.

Velikova V, Várkonyi Z, Szabó M, Maslenkova L, Nogues I, Kovács L, Peeva V, Busheva M, Garab G, Sharkey TD, Loreto F (2011) Increased thermostability of thylakoid membranes in isoprene-emitting leaves probed with three biophysical techniques. Plant Physiol. 157: 905–916.

Vile D, Pervent M, Belluau M, Vasseur F, Bresson J, Muller B, Granier C, Simonneau T (2012) *Arabidopsis* growth under prolonged high temperature and water deficit: Independent or interactive effects? Plant Cell Environ. 35: 702–718.

Wall JK, Johnson CB (1982) The effect of temperature on phytochrome controlled hypocotyl extension in *Sinapis alba* L. New Phytol. 405–412.

Wang LC, Tsai MC, Chang KY, Fan YS, Yeh CH, Wu SJ (2011) Involvement of the *Arabidopsis* HIT1/AtVPS53 tethering protein homologue in the acclimation of the plasma membrane to heat stress. J. Exp. Bot. 62: 3609–3620.

Walther G-R (2003) Plants in a warmer world. Perspect. Plant Ecol. Evol. Syst. 6 (3): 169–185.

Watt MS, Bloomberg M (2012) Key features of the seed germination response to high temperatures. New Phytol. 196: 332–336.

Went FW (1944) Plant growth under controlled conditions II. Thermoperiodicity in growth and fruiting of the tomato. Am. J. Bot. 31: 135–150.

Whittle C, Otto S, Johnston M, Krochko J (2009) Adaptive epigenetic memory of ancestral temperature regime in *Arabidopsis thaliana*. Botany 87: 650–657.

Willis CG, Ruhfel B, Primack RB, Miller-Rushing AJ, Davis CC (2008) Phylogenetic patterns of species loss in Thoreau's woods are driven by climate change. Proc. Natl. Acad. Sci. USA 105: 17029–17033.

Wu S-J, Locy RD, Shaw JJ, Cherry JH, Singh NK (2000) Mutation in *Arabidopsis* HIT1 locus causing heat and osmotic hypersensitivity. J. Plant Physiol. 157: 543–547.

Yamada M, Greenham K, Prigge MJ, Jensen PJ, Estelle M (2009). The TRANSPORT INHIBITOR RESPONSE2 gene is required for auxin synthesis and diverse aspects of plant development. Plant Physiol. 151: 168–179.

Yamori W, Noguchi K, Hikosaka K, Terashima I (2009) Cold-tolerant crop species have greater temperature homeostasis of leaf respiration and photosynthesis than cold-sensitive species. Plant Cell Physiol. 50: 203–215.

Yant L, Mathieu J, Schmid M (2009) Just say no: Floral repressors help *Arabidopsis* bide the time. Curr. Opin. Plant Biol. 12: 580–586.

Yu F, Berg VS (1994) Control of paraheliotropism in two phaseolus species. Plant Physiol. 106: 1567–1573.

4 Vernalization: Competence to flower provided by winter

Dong-Hwan Kim and Sibum Sung

4.1 Introduction

Floral transition is a critical developmental change in the life cycle of flowering plants. Plants optimize their flowering time in response to environmental cues to ensure reproductive success. In temperate climates, temperature fluctuation serves as an external stimulus to signal seasonal changes. Notably, extended periods of low temperature (i.e., winter) render plants to competent to flower the following spring. This phenomenon is known as vernalization. To ensure that plants promote floral transition during the correct time of year, multiple regulatory networks have evolved to sense seasonal fluctuations of temperature and day length. Our current knowledge of flowering time regulation and vernalization is mainly derived from studies using the model *Arabidopsis thaliana*. In *Arabidopsis*, four major flowering pathways have been defined by molecular genetic studies, namely, the autonomous pathway, the photoperiod pathway, the gibberellin (GA) pathway, and the vernalization pathway (Figure 4.1). In recent years, other temperature-responsive floral pathways have become evident. These are discussed in Chapter 3. Favorable conditions for flowering are translated into the activation of a series of downstream genes, such as *FLOWERING LOCUS T* (*FT*), *SUPPRESSOR OF OVEREXPRESSION OF CONSTANS 1* (*SOC1*), and *AGAMOUS-LIKE 24* (*AGL24*) (often referred to as floral integrators), which in turn induce floral transition (Moon et al. 2005; Kim et al. 2009). Two major flowering genes, *CONSTANS* (*CO*) and *FLOWERING LOCUS C* (*FLC*), act antagonistically on floral integrators, whereas GA acts directly to activate *SOC1* and *LEAFY* (*LFY*) (Figure 4.1). Both autonomous and vernalization pathways act in parallel to repress *FLC*. Mutations in autonomous pathway genes do not impair the vernalization response. In other words, increased *FLC* transcription, and thus late flowering, observed in autonomous pathway mutants can be fully suppressed by vernalization treatment. We will first describe the molecular basis for elevated *FLC* transcription observed in natural accessions of

Temperature and Plant Development, First Edition. Edited by Keara A. Franklin and Philip A. Wigge.

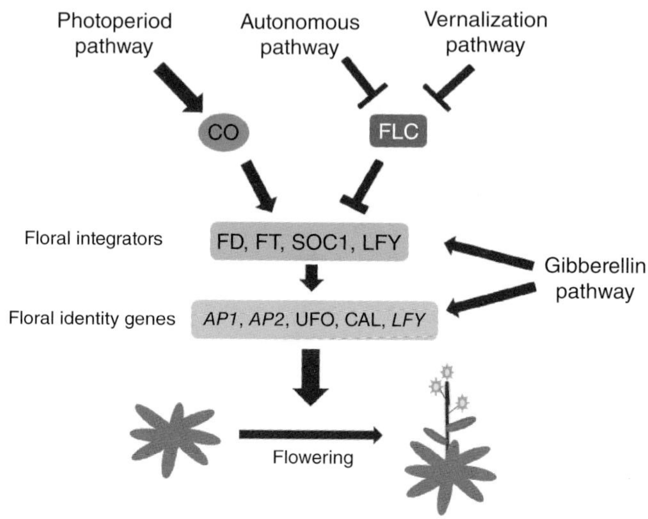

Figure 4.1 Four major flowering pathways in *Arabidopsis*. Vernalization and autonomous pathways commonly act to repress *FLC*. *FLC* acts to repress a group of floral integrators, including FT and SOC1, to block rapid flowering. The photoperiod pathway acts to activate floral integrators under inductive photoperiods. Activated floral integrators activate floral identity genes, such as *APETALA 1* (*AP1*), *APETALA 2* (*AP2*), *SEPALATA 3* (*SEP3*), *FRUITFUL* (*FUL*), and *LFY* to initiate flower development. GA-mediated pathway acts directly to downstream floral activators, including *SOC1* and *LFY*. For color detail, please see color plate section.

Arabidopsis and autonomous pathway mutants. Our current understanding of the molecular mechanisms of *Arabidopsis* vernalization will then be summarized. We will conclude by briefly discussing the diverse regulatory circuits that govern vernalization in other flowering plants.

4.2 Vernalization requirement in *Arabidopsis*

In temperate climates, some plants use winter as an indicator of seasonal change. In these plants, exposure to winter cold results in accelerated flowering (known as vernalization). Biennial plants cannot flower without prior exposure to winter cold (termed an obligate vernalization requirement), whereas winter-annual plants still flower, although late, without vernalizing cold (termed a facultative vernalization requirement). Vernalization is not, however, sufficient to induce flowering by itself. Rather, vernalized plants still need to be exposed to inductive photoperiods. Most plants with a vernalization requirement are also long-day (LD) plants. For these individuals, floral transition is triggered under long photoperiods in spring following the winter cold.

Most lab strains of *Arabidopsis* are typical summer annuals and do not require vernalization to achieve rapid flowering. The rapid-flowering *Arabidopsis* accessions (e.g., Columbia and Landsberg) have been popular for lab use because they flower and complete their life cycle quickly. A dominant locus, *FRIGIDA* (*FRI*), responsible for conferring vernalization requirement, was first identified from a cross between winter-annual and summer-annual strains of *Arabidopsis* (Napp-Zinn 1987). Subsequent genetic analysis demonstrated that *FRI* accounts for the winter-annual behavior in many *Arabidopsis* accessions (Burn et al. 1993; Lee et al. 1993; Clarke and Dean 1994; Koornneef et al. 1994). Genetic studies further revealed that another dominant allele, *FLC*, is required for *FRI* to confer winter-annual habit (Koornneef et al. 1994; Lee et al. 1994b). *FLC* encodes a MADS-box domain-containing DNA-binding protein and acts as a potent repressor of flowering (Michaels and Amasino 1999; Sheldon et al. 1999). Constitutive expression of *FLC* is sufficient to delay flowering in summer-annual strains of *Arabidopsis*. *FRI* functions to upregulate *FLC* expression to a level that inhibits floral transition (Michaels and Amasino 1999). Vernalization requirement in *Arabidopsis* is therefore mainly due to high levels of *FLC* transcription.

4.2.1 Molecular basis of FRI-*mediated* FLC *activation*

FRI encodes a novel plant-specific protein (Johanson et al. 2000). Suppressor screens to uncover components necessary for *FRI* function revealed a number of genes, including the related genes, *FRI-LIKE 1* (*FRL1*) and *FRI-LIKE 2* (*FRL2*). Other components of *FRI* signaling include FES1, SUF4, and FLX/SUF5 (Michaels et al. 2004; Schmitz et al. 2005; Kim et al. 2006; Andersson et al. 2008). Mutations in these components commonly result in early flowering despite the presence of an active *FRI* allele. They have little effect on *FLC* expression caused by mutations in autonomous pathway genes, suggesting that they are specific to *FRI*. The upregulation of *FLC* by FRI is therefore mediated by a mechanism separate from the autonomous pathway.

Biochemical purification of the FRI-containing complex has shed light on the molecular basis of *FLC* activation (Choi et al. 2011). One FRI-containing complex component, SUF4, is a BED-type zinc finger protein. SUF4 specifically recognizes and binds to a minimal 15 bp-sequence motif (-CCAAATTTTAAGTTT-) of the *FLC* promoter region. Recruitment of the FRI-containing complex appears to be directed by sequence-specific DNA-binding activity of SUF4. In addition, other activating chromatin-remodeling complex components co-purify with the core FRI-containing complex, including the SWR1 chromatin-remodeling complex and the ASH2-like SET domain protein, EARLY FLOWERING IN SHORT

DAYS (EFS; also known as SDG8, ASHH2, or CCR1). Both the SWR1 chromatin-remodeling complex and EFS also function to activate *FLC* transcription in autonomous pathway mutants. The FRI-containing complex therefore utilizes universal transcriptional activation machineries to achieve high levels of *FLC* transcription. Mutations in *EFS* result in pleiotropic effects on other developmental processes, including early flowering, reduced organ size, enhanced shoot branching, altered carotenoid composition, and reduced fertility (Kim et al. 2005; Zhao et al. 2005; Xu et al. 2008; Ko et al. 2010). EFS has a dual function to mediate di- and trimethylation of both histone H3 Lys 4 (H3K4) and H3 Lys 36 (H3K36) at *FLC* chromatin (Kim et al. 2005; Zhao et al. 2005; Xu et al. 2008; Ko et al. 2010). The FRI-containing complex functions to enhance the activity of EFS, resulting in higher levels of H3K4 and H3K36 methylation (Ko et al. 2010).

4.2.2 Mutations in autonomous pathway genes: Another route to confer vernalization requirement

Late-flowering mutants in summer-annual strains of *Arabidopsis* can be classified into two groups. In one group, flowering is delayed under LD but is comparable to the wild type under short days (SD). This group of mutants is blind to inductive photoperiod (i.e., LD) and thus referred to as photoperiod mutants. The other group contains mutants which flower later than the wild type both under LD and SD and are referred to as autonomous pathway mutants (Koornneef et al. 1991). A characteristic of autonomous pathway mutants is that their late-flowering phenotype is fully suppressed by vernalization (Koornneef et al. 1991). Flowering behavior under different photoperiods and vernalization responsiveness can together therefore be used to classify late-flowering mutants.

The molecular cloning of *FLC* and subsequent genetic studies have revealed that late flowering in autonomous pathway mutants depends entirely on the presence of *FLC* (Michaels and Amasino 1999, 2001). Autonomous pathway genes include *LUMINIDEPENDENS* (*LD*), *FCA*, *FPA*, *FY*, *FLOWERING LOCUS D* (*FLD*), *FVE*, *FLOWERING LOCUS K* (*FLK*), and *RELATIVE OF EARLY FLOWERING6* (*REF6*) (Koornneef et al. 1991; Lee et al. 1994a; Macknight et al. 1997; Schomburg et al. 2001; He et al. 2003; Simpson et al. 2003; Ausin et al. 2004; Lim et al. 2004; Mockler et al. 2004; Noh et al. 2004). *FCA*, *FPA*, and *FLK* encode RNA-binding domain-containing proteins, and FY is homologous to a yeast polyadenylation factor 1 subunit 2 (Pfs2p), which functions in 3′ end cleavage and polyadenylation of mRNA. Other autonomous pathway genes, *FLD* and *REF6*, encode two different types of histone demethylase (He et al. 2003; Noh et al. 2004; Jiang et al. 2007). The biochemical properties of these autonomous pathway proteins led to a model in which

FLC repression by the autonomous pathway involves RNA-processing (e.g., RNA polyadenylation) and chromatin-modifying events (Michaels 2009). This hypothesis was reinforced by the identification of *fld* as a suppressor of *FCA* overexpression-mediated early flowering (Liu et al. 2007). Although both *FCA* and *FPA* were shown to be responsible for alternative processing of their own RNA transcripts, mutations in *FCA* and *FY* do not result in alternative processing of *FLC* mRNA (Quesada et al. 2003; Hornyik et al. 2010).

4.2.3 Other chromatin-remodeling complexes required for FLC *activation*

Genetic screens have identified another group of nuclear proteins required for *FLC* activation both in the presence of active *FRI* allele and in autonomous pathway mutants (Reeves et al. 2002; Zhang and van Nocker 2002; Noh and Amasino 2003; He et al. 2004; Oh et al. 2004; He and Amasino 2005; Kim et al. 2005; Choi et al. 2007). These proteins are related to evolutionarily conserved activating chromatin-remodeling factors and include homologues to components of Trithorax group (TrxG) complexes. Increased levels of methylation at H3K4 and H3K36 are prominent histone marks at FLC chromatin in both *FRI*-containing accessions and autonomous pathway mutants. This second group of proteins is commonly required for the deposition of similar histone marks at *FLC* and other loci. For instance, *EFS* encodes a SET2/Ash1L-type H3K36 methyltransferase and is required for H3K36 and H3K4 methylation at *FLC* chromatin (Kim et al. 2005; Ko et al. 2010). ATX1 and its related proteins, ATX2, ATXR3, and ATXR7, act redundantly as SET1/MLL-type methyltransferases as components of the COMPASS-like complex and are necessary for H3K4 methylation at many loci including *FLC* (Saleh et al. 2008; Tamada et al. 2009; Yun et al. 2012). Components of the RNA polymerase II-associated factor 1 (PAF1) complex, *ELF7* and *ELF8*, also act coordinately to ensure activation of *FLC* (He et al. 2004; He 2012). A recent study has shown that a COMPASS-like complex functions to establish an active RNA polymerase II transcription initiation complex (PIC) and its H3K4 methylation activity is required for efficient elongation of the polymerase in *Arabidopsis* (Ding et al. 2012). ATX1 is further required for the recruitment of another COMPASS-like complex component, WDR5, to PIC and interacts with general transcriptional machineries (Ding et al. 2012). The FRI-containing complex tightly associates with some of these proteins, as evident from biochemical purifications and co-immunoprecipitation studies (Jiang et al. 2009; Choi et al. 2011). Together, they directly enhance transcriptional activity in the presence of FRI-containing complexes at *FLC* chromatin. It remains to be addressed how the autonomous pathway proteins coordinate these common activating chromatin complexes.

4.3 The molecular mechanism of vernalization

The first clue as to the molecular basis of vernalization came from the cloning of *FLC* (Michaels and Amasino 1999; Sheldon et al. 1999). Vernalizing cold triggers repression of *FLC* which is then stably maintained. The repressed state of *FLC* persists throughout subsequent mitotic cell divisions when plants return to warm growth temperature, indicating an epigenetic mechanism. Prior to winter, FLC directly binds to the promoter regions of several key floral integrators, such as FT, FD, and SOC1, inhibiting their transcription (Figure 4.2; Helliwell et al. 2006; Searle et al. 2006). Vernalization-mediated *FLC* repression relieves the inhibitory effect of *FLC* on these floral integrators. As a result, plants become competent to flower. In the absence of *FLC*, inductive photoperiods can activate floral integrators (Figure 4.2).

4.3.1 Vernalization-mediated epigenetic repression of FLC

Genetic screens to uncover mutants with a defective vernalization response have revealed genes involved in the stable repression of *FLC*, including *VERNALIZATION1* (*VRN1*), *VERNALIZATION2* (*VRN2*), and *VERNALIZATION INSENSITIVE3* (*VIN3*). *VRN1* encodes a plant-specific B3 DNA-binding protein (Chandler et al. 1996; Gendall et al. 2001; Levy et al. 2002;

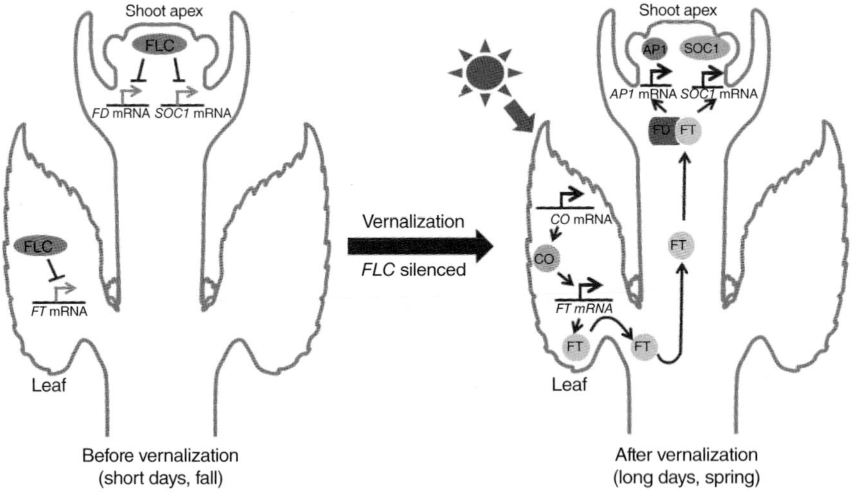

Figure 4.2 Vernalization-mediated flowering in *Arabidopsis*. Prior to vernalization, high levels of FLC prevent *FT*, *FD*, and *SOC1* expression and flowering is inhibited. After vernalization, *FLC* is silenced. In the absence of a prohibitory effect by FLC, floral integrators, FT, FD, and SOC1, can be activated by inductive photoperiods. FT protein has been shown to be a mobile floral activator which initiates flowering. For color detail, please see color plate section.

Sung and Amasino 2004). *VRN2* encodes an *Arabidopsis* homologue of SUPPRESSOR OF ZESTE 12 (SUZ12), which is a core component of Polycomb Repression Complex 2 (PRC2) (Gendall et al. 2001). VIN3 encodes a Plant Homeodomain (PHD) finger – containing protein which biochemically co-purifies with core components of PRC2 (Wood et al. 2006; De Lucia et al. 2008). In contrast to VRN1 and VRN2, which are constitutively expressed, VIN3 is only expressed during the cold exposure (Gendall et al. 2001; Levy et al. 2002; Sung and Amasino 2004). For VIN3 during and after vernalization, the levels of certain histone modifications associated with gene activation are reduced at *FLC* chromatin, such as acetylation at histone H3 Lys 9 (H3K9) and histone H3 Lys 14 (H3K14) and methylation at H3K4 (Bastow et al. 2004; Sung and Amasino 2004). By contrast, two repressive histone modifications (methylation at H3K9 and H3K27) increase (Bastow et al. 2004; Sung and Amasino 2004; Sung et al. 2006b; Greb et al. 2007). Elevated levels of H3K9 and H3K27 methylation are typically associated with stably repressed chromatin. VIN3 preferentially binds to dimethylated H3K9 (H3K9me2) peptides *in vitro* and is required for the PRC2 activity – trimethylation of H3K27 (H3K27me3) (Kim and Sung 2013). Enrichments of both VIN3 and PRC2 increase at *FLC* chromatin as a result of vernalization (Sung and Amasino 2004; Kim and Sung 2013). It therefore appears that VIN3 reinforces PRC2 activity.

PRC2 components are very well conserved in *Arabidopsis* (Hsieh et al. 2003). These include the H3K27 methyltransferase, Enhancer of Zeste (E(z)). In *Arabidopsis*, CURLY LEAF (CLF), SWINGER (SWN), and MEDEA (MEA) are homologues of E(z). MEA functions only in the endosperm and CLF and SWN function in sporophytes. In *clf* and *swn* mutants, *FLC* is derepressed and the level of H3K27me3 at *FLC* chromatin is compromised compared to the wild type, indicating that CLF and SWN are components of PRC2 responsible for the *FLC* repression. Indeed, both CLF and SWN co-purify with the VIN3-containing PRC2 complex (Wood et al. 2006; De Lucia et al. 2008). In many higher eukaryotes, H3K27me3 histone marks deposited by PRC2 on target chromatin are recognized by another complex, known as Polycomb Repression Complex 1 (PRC1) (Schwartz and Pirrotta 2007). Although there are no apparent homologues of some PRC1 components in *Arabidopsis*, the existence of a PRC1-like complex is supported by several studies. LIKE-HETEROCHROMATIN PROTEIN 1 (LHP1), an *Arabidopsis* homologue of HP1, is a chromodomain-containing protein which specifically recognizes and binds to H3K27me3 at target chromatin (Mylne et al. 2006; Sung et al. 2006a). EMBRYONIC FLOWERING 1 (EMF1) appears to function as a PRC1-like component (Calonje et al. 2008). In addition, two related RING-finger proteins, AtBMI1A and AtBMI1B, interact with both EMF1 and LHP1. AtBMI1A and AtBMI1B mediate histone H2A monoubiquitination, which is a known

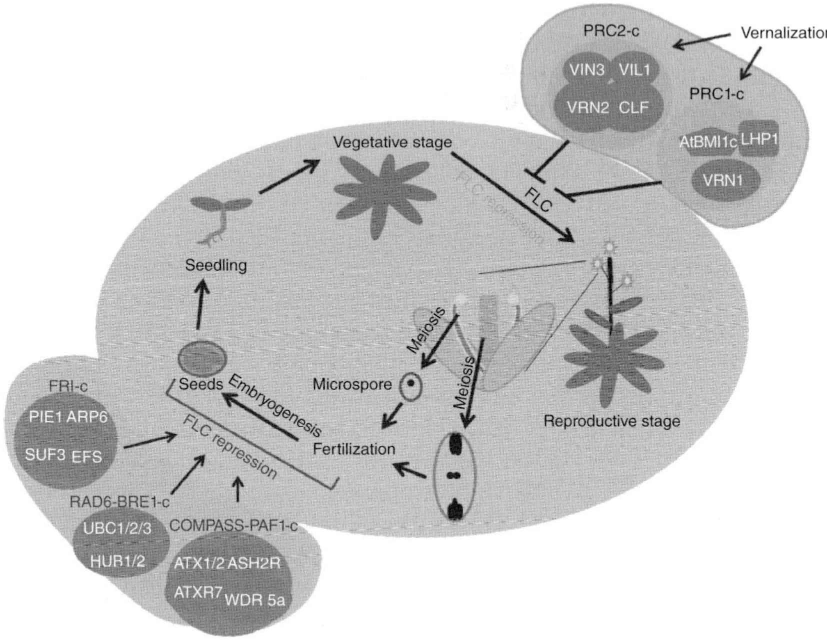

Figure 4.3 Fates of *FLC* chromatin during the *Arabidopsis* life cycle. Repressive complexes, PRC2 and PRC1, cause mitotically stable repression of *FLC* by vernalization. After meiosis, repressed *FLC* is reactivated by activating chromatin-remodeling complexes, including the FRIGIDA-containing complex (FRI-c), RAD6/BRE1-containing complex (RAD6-BRE1-c), and COMPASS/PAF1-containing complex (COMPASS-PAF1-c). For color detail, please see color plate section.

biochemical activity of PRC1 in other eukaryotes (Bratzel et al. 2010). The PRC1-like complex in *Arabidopsis* therefore contains at least LHP1, EMF1, AtBMI1A, and AtBMI1B (Bratzel et al. 2010). Another RING-finger protein, AtBMI1C, was also shown to be involved in repression of *FLC*. Similar to AtBMI1 and AtBMI2, AtBMI1C mediates monoubiquitination of histone H2A (H2Aub1) at *FLC* by vernalization (Li et al. 2011). Evolutionarily conserved PRC2 and PRC1 therefore contribute to the stable repression of *FLC* in response to vernalization in *Arabidopsis* (Figure 4.3).

4.3.2 The dynamics of PRC2 and TRX at FLC *chromatin*

Polycomb and Trithorax complexes compete at the same target genes. TrxG mediates methylation at H3K4 when *FLC* is actively transcribed (before winter), whereas PRC2 mediates methylation at H3K27 when *FLC* is repressed (during and after winter). How *FLC* chromatin transits from

TrxG-dominant chromatin to PRC2-dominant chromatin is not clear. There are several possibilities as to how *FLC* becomes PRC2-dominant chromatin by vernalization: (a) TrxG is constantly associated with *FLC* chromatin and enrichments of PRC2 increase with vernalization; (b) TrxG is excluded from *FLC* chromatin, while PRC2 remains constant during vernalization; and (c) TrxG is excluded from *FLC* chromatin and enrichment of PRC2 increases during vernalization. Chromatin-immunoprecipitation (ChIP) assays used to trace TrxG and PRC2 components suggest that TrxG is excluded and PRC2 is recruited to *FLC* chromatin by vernalization (Kim and Sung 2013). The net result of these TrxG and PRC2 dynamics is increased H3K27 methylation and decreased H3K4 methylation at *FLC* chromatin. It remains to be determined whether TrxG is outcompeted by PRC2 (i.e., an active mechanism) or TrxG is diluted out (i.e., a passive mechanism) as a result of vernalization.

4.3.3 Mechanisms underlying PRC2 recruitment to FLC *chromatin by vernalization*

Most components of PRC2 are constitutively expressed and do not show vernalization-specific expression patterns. VIN3 is the only known protein component that is cold specific. VIN3 and other known PRC2 components do not, however, possess sequence-specific DNA-binding activity that can be used to direct PRC2 specifically to *FLC* chromatin upon vernalization. How PRC2 achieves selective gene repression in a locus-specific manner is not well understood. In recent years, there have been a number of reports identifying regulatory RNAs, some of which act to direct chromatin-remodeling activities to certain target loci *in cis* and/or *in trans* (Tsai et al. 2010; Nagano and Fraser 2011). Examples include long noncoding RNAs (lncRNAs) that direct repressive activity, including PRC2, to target loci. Several lncRNAs, including *ANRIL*, *HOTAIR*, *Xist*, and *Kcnq1ot1*, physically interact with PRC2 components and are required for locus-specific gene repression in mammals (Kotake et al. 2003; Pandey et al. 2008; Zhao et al. 2008; Tsai et al. 2010). In addition, many lncRNAs have copurified with various components of chromatin-modifying complexes, suggesting that lncRNAs may serve as a scaffold (Tsai et al. 2010; Guttman et al. 2011; Spitale et al. 2011).

In vernalization-mediated *FLC* repression, a group of lncRNAs, collectively named *COOLAIR*, have been identified from the *FLC* genomic region (Swiezewski et al. 2009). They are transcribed in an antisense direction and increase transiently during vernalization. Two classes of *COOLAIR* antisense RNAs are expressed, based on transcript size. Alternative 3′ end processing has been implicated to have a role in autonomous pathway-mediated *FLC* repression (Liu et al. 2010). Longer *COOLAIR* transcripts start from the 3′ noncoding region and extend beyond the transcriptional start site of sense *FLC* transcripts. Loss-of-function alleles of *COOLAIR* do not, however, compromise either flowering time or vernalization response (Helliwell et al. 2011). The molecular basis of *COOLAIR* function in vernalization therefore remains to be determined.

In the autonomous flowering pathway, alternative 3′ processing of *COOLAIR* has been proposed to mediate *FLC* repression. *COOLAIR* transcripts can be differentially polyadenylated and thus grouped into either proximal transcripts (end within the 6th exon of sense *FLC*) or distal transcripts (end at the promoter region of sense *FLC*). It was speculated that *COOLAIR* transcripts interact with FCA through its RRM domains. This mechanism promotes usage of the proximal poly(A) site over the distal poly(A) site. Observations that the relative usage of the distal poly(A) site increases in *fca*, *fpa*, or *fy* mutants led to the suggestion that increased antisense *COOLAIR* transcription through the sense *FLC* promoter may stimulate sense transcription (Liu et al. 2010; Ietswaart et al. 2012). In this model, the choice of 3′ end processing between proximal poly(A) and distal poly(A) sites could antagonistically function as either a repressor or an activator of sense *FLC* transcription in a co-transcriptional manner (Crevillen and Dean 2010). How *COOLAIR* processing functions in vernalization is not clear. Autonomous pathway mutants, including *fca* and *fy* in which transcription and 3′ end processing of *COOLAIR* are altered, do not show any defect in vernalization response.

By employing a tiled 5′ RACE followed by RT-PCR during the course of vernalization, another lncRNA, *COLD ASSISTED INTRONIC NONCODING RNA* (*COLDAIR*), was identified (Heo and Sung 2011). *COLDAIR* is transcribed from the first intron of *FLC* in a sense direction and shows a transiently induced expression pattern during the course of vernalization. *COLDAIR* transcript (~1.1 KB long) physically interacts with CLF, one of the PRC2 components, both *in vitro* and *in vivo*. Knockdown of *COLDAIR* using RNAi compromises vernalization-mediated PRC2 recruitment at *FLC* chromatin and thus results in reduced H3K27 methylation. *COLDAIR* is therefore at least part of a mechanism to direct PRC2 to *FLC* chromatin during vernalization.

4.4 Resetting of *FLC* repression during meiosis

Vernalization-mediated *FLC* repression is stably maintained throughout mitotic cell division even after plants return to warm temperature. A vernalization requirement is, however, restored in the next generation. This is an important characteristic of the vernalization response; if vernalization-mediated *FLC* repression was stable after meiosis, no plant species would require vernalization after one round of winter season. *FLC* appears to be reactivated during early embryogenesis after fertilization (Choi et al. 2009). *FRI*-dependent *FLC* activation is observed, and this requires some activating chromatin-remodeling complex components, including PHOTOPERIOD-INDEPENDENT EARLY FLOWERING1 (PIE1). Activating chromatin-remodeling complexes therefore occupy *FLC* chromatin at this stage (Figure 4.3). Repressed *FLC* chromatin must have been reprogrammed either during meiosis or at an earlier stage of embryogenesis. How this epigenetic reprogramming is achieved at *FLC* chromatin remains to be determined.

4.5 Vernalization in other plant species

Although much of our understanding of vernalization has been gained from *Arabidopsis* research, studies using other species show interesting similarities as well as differences in molecular components. It appears that one common theme of the vernalization response is 'repression of floral repressor'. It should, however, be noted that the presence of *FLC* is not apparent outside of Brassicaceae. Regulatory circuits underlying both vernalization requirement and response may therefore have evolved independently among flowering plants. Here we summarize a few examples of evolutionarily diverse vernalization regulatory circuits in other species.

4.5.1 Arabis alpina

Arabis alpina, a close relative of *Arabidopsis*, has a distinctive vernalization response (Koch et al. 2006; Ansell et al. 2008). Unlike *Arabidopsis*, which is an annual plant, *Arabis* is a perennial plant and repeats cycles of vegetative and reproductive development. In annual plants, all apical meristems initiate floral development at the same time in their life cycle, termed monocarpy. In contrast, perennial plants are polycarpic and sustain vegetative development after flowering, which allows plants to bloom and set seed many times in their life cycle. Mutant screens to identify genes conferring vernalization requirement in perennial *Arabis* identified *PERPETUAL FLOWERING 1* (*PEP1*), an *Arabis* orthologue of *FLC* (Wang et al. 2009). Similar to *Arabidopsis FLC*, *PEP1* is downregulated in response to vernalization. The repression of *PEP1* is, however, only stable during cold exposure, and *PEP1* expression is eventually reactivated after plants are returned to warm growth temperature. Consistent with the unstable repression of *PEP1*, a repressive histone mark, H3K27me3, is enriched at *PEP1* chromatin during cold exposure but decreased once plants return to warm temperature. This unstable nature of *PEP1* repression is likely responsible for the polycarpic flowering behavior in perennial *Arabis*.

4.5.2 Cereals (wheat and barley)

In temperate cereals, such as wheat and barley, winter cultivars exhibit a vernalization requirement for rapid flowering, whereas spring cultivars flower rapidly without vernalization. Genetic studies between winter and spring cultivars have identified several of the genes involved (Trevaskis et al. 2007; Distelfeld et al. 2009). Three loci, *VRN1*, *VRN2*, and *VRN3* (which are not related with *Arabidopsis VRN1* and *VRN2*), are responsible for conferring a vernalization requirement in temperate cereals (Yan et al. 2003; Yan et al. 2004; Yan et al. 2006; Trevaskis et al. 2007; Distelfeld et al. 2009). There is no apparent orthologue of *FLC* in cereals. Instead, VRN2 is a potent floral repressor. VRN2 contains a zinc finger motif and a CONSTANS, CONSTANS-LIKE, and TIMING OF CAB1-1 (CCT) domain (Yan et al. 2004). *VRN2* acts to block flowering and is repressed by vernalization. In a simple model, *VRN2* can be considered as a functional homologue of *FLC*. Unlike *FLC*, however, *VRN2* is also under the control of photoperiods. There are therefore significant differences in regulatory circuits between *Arabidopsis* and cereals to confer vernalization requirement and response.

*4.5.3 Sugar beet (*Beta vulgaris*)*

Sugar beet contains an *FLC* orthologue, *BvFL1*, which, when ectopically expressed in *Arabidopsis*, confers a late-flowering phenotype. *BvFL1* is modestly downregulated in response to vernalization (Reeves et al. 2007). The repression of *BvFL1* is, however, not stable when plants are returned to warm growth temperature. *BvFL1* may therefore possess only limited floral repressor activity in sugar beet. Using a reverse genetic approach, a pair of FT homologues was identified as major sugar beet floral regulators (Pin et al. 2010). Interestingly, these two very closely related genes act oppositely in flowering regulation. Bv*FT1* acts as a floral repressor (in a similar manner to *Arabidopsis FLC*) and is stably repressed by vernalization, while its homologue, Bv*FT2*, is required for the promotion of flowering (in a similar manner to *Arabidopsis FT*) (Pin et al. 2010). It therefore appears that sugar beet uses alternative regulatory components to mediate vernalization.

Despite differences in the components of regulatory circuits, fundamental vernalization physiology is similar between different species. Repression of a floral repressor by an extended period of cold renders plants competent to flower the following spring. Whether different floral repressors are also under the control of chromatin modifications remains to be addressed.

4.6 Concluding remarks

Vernalization studies in the model plant *Arabidopsis* have provided us with the molecular basis of the response. Given the diversity of flowering plants and their evolutionary origins, variations in the regulatory circuits used to control vernalization are expected. Further studies of vernalization in *Arabidopsis* and other plant species will provide valuable insight into how plants use temperature fluctuation to determine their developmental program.

Acknowledgments

We apologize that we could not include some of the wonderful works in vernalization due to our focused discussion on the topic. Our work on vernalization is currently supported by NSF and NIH.

References

Andersson CR, Helliwell CA, Bagnall DJ, Hughes TP, Finnegan EJ, Peacock WJ, Dennis ES (2008) The FLX gene of *Arabidopsis* is required for FRI-dependent activation of FLC expression. Plant Cell Physiol. 49: 191–200.

Ansell SW, Grundmann M, Russell SJ, Schneider H, Vogel JC (2008) Genetic discontinuity, breeding-system change and population history of *Arabis alpina* in the Italian Peninsula and adjacent Alps. Mol. Ecol. 17: 2245–2257.

Ausin I, Alonso-Blanco C, Jarillo JA, Ruiz-Garcia L, Martinez-Zapater JM (2004) Regulation of flowering time by FVE, a retinoblastoma-associated protein. Nat. Genet. 36: 162–166.

Bastow R, Mylne JS, Lister C, Lippman Z, Martienssen RA, Dean C (2004) Vernalization requires epigenetic silencing of FLC by histone methylation. Nature 427: 164–167.

Bratzel F, Lopez-Torrejon G, Koch M, Del Pozo JC, Calonje M (2010) Keeping cell identity in *Arabidopsis* requires PRC1 RING-finger homologs that catalyze H2A monoubiquitination. Curr. Biol. 20: 1853–1859.

Burn JE, Bagnall DJ, Metzger JD, Dennis ES, Peacock WJ (1993) DNA methylation, vernalization, and the initiation of flowering. Proc. Natl. Acad. Sci. USA 90: 287–291.

Calonje M, Sanchez R, Chen L, Sung ZR (2008) EMBRYONIC FLOWER1 participates in polycomb group-mediated AG gene silencing in *Arabidopsis*. Plant Cell 20: 277–291.

Chandler J, Wilson A, Dean C (1996) *Arabidopsis* mutants showing an altered response to vernalization. Plant J. 10: 637–644.

Choi K, Park C, Lee J, Oh M, Noh B, Lee I (2007) *Arabidopsis* homologs of components of the SWR1 complex regulate flowering and plant development. Development 134: 1931–1941.

Choi J, Hyun Y, Kang MJ, In Yun H, Yun JY, Lister C, Dean C, Amasino RM, Noh B, Noh YS, Choi Y (2009) Resetting and regulation of FLOWERING LOCUS C expression during *Arabidopsis* reproductive development. Plant J. 57: 918–931.

Choi K, Kim J, Hwang HJ, Kim S, Park C, Kim SY, Lee I (2011) The FRIGIDA complex activates transcription of FLC, a strong flowering repressor in Arabidopsis, by recruiting chromatin modification factors. Plant Cell 23: 289–303.

Clarke JH, Dean C (1994) Mapping FRI, a locus controlling flowering time and vernalization response in *Arabidopsis thaliana*. Mol. Gen. Genet. 242: 81–89.

Crevillen P, Dean C (2010) Regulation of the floral repressor gene FLC: The complexity of transcription in a chromatin context. Curr. Opin. Plant Biol. 14: 38–44.

De Lucia F, Crevillen P, Jones AM, Greb T, Dean C (2008) A PHD-polycomb repressive complex 2 triggers the epigenetic silencing of FLC during vernalization. Proc. Natl. Acad. Sci. USA 105: 16831–16836.

Ding Y, Ndamukong I, Xu Z, Lapko H, Fromm M, Avramova Z (2012) ATX1-generated H3K4me3 is required for efficient elongation of transcription, not initiation, at ATX1-regulated genes. PLoS Genet. 8: e1003111.

Distelfeld A, Li C, Dubcovsky J (2009) Regulation of flowering in temperate cereals. Curr Opin. Plant Biol. 12: 178–184.

Gendall AR, Levy YY, Wilson A, Dean C (2001) The VERNALIZATION 2 gene mediates the epigenetic regulation of vernalization in *Arabidopsis*. Cell 107: 525–535.

Greb T, Mylne JS, Crevillen P, Geraldo N, An H, Gendall AR, Dean C (2007) The PHD finger protein VRN5 functions in the epigenetic silencing of *Arabidopsis* FLC. Curr. Biol. 17: 73–78.

Guttman M, Donaghey J, Carey BW, Garber M, Grenier JK, Munson G, Young G, Lucas AB, Ach R, Bruhn L, Yang X, Amit I, Meissner A, Regev A, Rinn JL, Root DE, Lander ES (2011) lincRNAs act in the circuitry controlling pluripotency and differentiation. Nature 477: 295–300.

He Y (2012) Chromatin regulation of flowering. Trends Plant Sci. 17: 556–562.

He Y, Amasino RM (2005) Role of chromatin modification in flowering-time control. Trends Plant Sci. 10: 30–35.

He Y, Michaels SD, Amasino RM (2003) Regulation of flowering time by histone acetylation in *Arabidopsis*. Science 302: 1751–1754.

He Y, Doyle MR, Amasino RM (2004) PAF1-complex-mediated histone methylation of FLOWERING LOCUS C chromatin is required for the vernalization-responsive, winter-annual habit in *Arabidopsis*. Genes. Dev. 18: 2774–2784.

Helliwell CA, Wood CC, Robertson M, James Peacock W, Dennis ES (2006) The *Arabidopsis* FLC protein interacts directly in vivo with SOC1 and FT chromatin and is part of a high-molecular-weight protein complex. Plant J. 46: 183–192.

Helliwell CA, Robertson M, Finnegan EJ, Buzas DM, Dennis ES (2011) Vernalization-repression of *Arabidopsis* FLC requires promoter sequences but not antisense transcripts. PLoS One 6: e21513.

Heo JB, Sung S (2011) Vernalization-mediated epigenetic silencing by a long intronic noncoding RNA. Science 331: 76–79.

Hornyik C, Terzi LC, Simpson GG (2010) The spen family protein FPA controls alternative cleavage and polyadenylation of RNA. Dev. Cell 18: 203–213.

Hsieh TF, Hakim O, Ohad N, Fischer RL (2003) From flour to flower: How Polycomb group proteins influence multiple aspects of plant development. Trends Plant Sci. 8: 439–445.

Ietswaart R, Wu Z, Dean C (2012) Flowering time control: Another window to the connection between antisense RNA and chromatin. Trends Genet. 28: 445–453.

Jiang D, Yang W, He Y, Amasino RM (2007) *Arabidopsis* relatives of the human lysine-specific Demethylase1 repress the expression of FWA and FLOWERING LOCUS C and thus promote the floral transition. Plant Cell 19: 2975–2987.

Jiang D, Gu X, He Y (2009) Establishment of the winter-annual growth habit via FRIGIDA-mediated histone methylation at FLOWERING LOCUS C in *Arabidopsis*. Plant Cell 21: 1733–1746.

Johanson U, West J, Lister C, Michaels S, Amasino R, Dean C (2000) Molecular analysis of FRIGIDA, a major determinant of natural variation in *Arabidopsis* flowering time. Science 290: 344–347.

Kim DH, Sung S (2013) Coordination of the vernalization response through a VIN3 and FLC gene family regulatory network in *Arabidopsis*. Plant Cell 25: 454–469.

Kim SY, He Y, Jacob Y, Noh YS, Michaels S, Amasino R (2005) Establishment of the vernalization-responsive, winter-annual habit in *Arabidopsis* requires a putative histone H3 methyl transferase. Plant Cell 17: 3301–3310.

Kim S, Choi K, Park C, Hwang HJ, Lee I (2006) SUPPRESSOR OF FRIGIDA4, encoding a C2H2-Type zinc finger protein, represses flowering by transcriptional activation of *Arabidopsis* FLOWERING LOCUS C. Plant Cell 18: 2985–2998.

Kim DH, Doyle MR, Sung S, Amasino RM (2009) Vernalization: Winter and the timing of flowering in plants. Annu. Rev. Cell Dev. Biol. 25: 277–299.

Ko JH, Mitina I, Tamada Y, Hyun Y, Choi Y, Amasino RM, Noh B, Noh YS (2010) Growth habit determination by the balance of histone methylation activities in *Arabidopsis*. EMBO. J. 29: 3208–3215.

Koch MA, Kiefer C, Ehrich D, Vogel J, Brochmann C, Mummenhoff K (2006) Three times out of Asia Minor: the phylogeography of *Arabis alpina* L. (Brassicaceae). Mol. Ecol. 15: 825–839.

Koornneef M, Hanhart CJ, van der Veen JH (1991) A genetic and physiological analysis of late flowering mutants in *Arabidopsis thaliana*. Mol. Gen. Genet. 229: 57–66.

Koornneef M, Blankestijn-de Vries H, Hanhart C, Soppe W, Peeters T (1994) The phenotype of some late-flowering mutants is enhanced by a locus on chromosome 5 that is not effective in the Landsberg erecta wild-type. Plant J. 6: 911–919.

Kotake T, Takada S, Nakahigashi K, Ohto M, Goto K (2003) *Arabidopsis* TERMINAL FLOWER 2 gene encodes a heterochromatin protein 1 homolog and represses both FLOWERING LOCUS T to regulate flowering time and several floral homeotic genes. Plant Cell Physiol. 44: 555–564.

Lee I, Bleecker A, Amasino R (1993) Analysis of naturally occurring late flowering in *Arabidopsis thaliana*. Mol. Gen. Genet. 237: 171–176.

Lee I, Aukerman MJ, Gore SL, Lohman KN, Michaels SD, Weaver LM, John MC, Feldmann KA, Amasino RM (1994a) Isolation of LUMINIDEPENDENS: A gene involved in the control of flowering time in *Arabidopsis*. Plant Cell 6: 75–83.

Lee I, Michaels SD, Masshardt AS, Amasino RM (1994b) The late-flowering phenotype of FRIGIDA and mutations in LUMINIDEPENDENS is suppressed in the Landsberg erecta strain of *Arabidopsis*. Plant J. 6: 903–909.

Levy YY, Mesnage S, Mylne JS, Gendall AR, Dean C (2002) Multiple roles of *Arabidopsis* VRN1 in vernalization and flowering time control. Science 297: 243–246.

Li W, Wang Z, Li J, Yang H, Cui S, Wang X, Ma L (2011) Overexpression of AtBMI1C, a polycomb group protein gene, accelerates flowering in *Arabidopsis*. PLoS One 6: e21364.

Lim MH, Kim J, Kim YS, Chung KS, Seo YH, Lee I, Kim J, Hong CB, Kim HJ, Park CM (2004) A new *Arabidopsis* gene, FLK, encodes an RNA binding protein with K homology motifs and regulates flowering time via FLOWERING LOCUS C. Plant Cell 16: 731–740.

Liu F, Quesada V, Crevillen P, Baurle I, Swiezewski S, Dean C (2007) The *Arabidopsis* RNA-binding protein FCA requires a lysine-specific demethylase 1 homolog to downregulate FLC. Mol. Cell 28: 398–407.

Liu F, Marquardt S, Lister C, Swiezewski S, Dean C (2010) Targeted 3′ processing of antisense transcripts triggers *Arabidopsis* FLC chromatin silencing. Science 327: 94–97.

Macknight R, Bancroft I, Page T, Lister C, Schmidt R, Love K, Westphal L, Murphy G, Sherson S, Cobbett C, Dean C (1997) FCA, a gene controlling flowering time in *Arabidopsis*, encodes a protein containing RNA-binding domains. Cell 89: 737–745.

Michaels SD (2009) Flowering time regulation produces much fruit. Curr. Opin. Plant Biol. 12: 75–80.

Michaels SD, Amasino RM (1999) FLOWERING LOCUS C encodes a novel MADS domain protein that acts as a repressor of flowering. Plant Cell 11: 949–956.

Michaels SD, Amasino RM (2001) Loss of FLOWERING LOCUS C activity eliminates the late-flowering phenotype of FRIGIDA and autonomous pathway mutations but not responsiveness to vernalization. Plant Cell 13: 935–941.

Michaels SD, Bezerra IC, Amasino RM (2004) FRIGIDA-related genes are required for the winter-annual habit in *Arabidopsis*. Proc. Natl. Acad. Sci. U S A 101: 3281–3285.

Mockler TC, Yu X, Shalitin D, Parikh D, Michael TP, Liou J, Huang J, Smith Z, Alonso JM, Ecker JR, Chory J, Lin C (2004) Regulation of flowering time in *Arabidopsis* by K homology domain proteins. Proc. Natl. Acad. Sci. USA 101: 12759–12764.

Moon J, Lee H, Kim M, Lee I (2005) Analysis of flowering pathway integrators in *Arabidopsis*. Plant Cell Physiol. 46: 292–299.

Mylne JS, Barrett L, Tessadori F, Mesnage S, Johnson L, Bernatavichute YV, Jacobsen SE, Fransz P, Dean C (2006) LHP1, the *Arabidopsis* homologue of HETEROCHROMATIN PROTEIN1, is required for epigenetic silencing of FLC. Proc. Natl. Acad. Sci. USA 103: 5012–5017.

Nagano T, Fraser P (2011) No-nonsense functions for long noncoding RNAs. Cell 145: 178–181.

Napp-Zinn K (ed.) (1987) Vernalization: Environmental and genetic regulation. Butterworths, London.

Noh YS, Amasino RM (2003) PIE1, an ISWI family gene, is required for FLC activation and floral repression in *Arabidopsis*. Plant Cell 15: 1671–1682.

Noh B, Lee SH, Kim HJ, Yi G, Shin EA, Lee M, Jung KJ, Doyle MR, Amasino RM, Noh YS (2004) Divergent roles of a pair of homologous jumonji/zinc-finger-class transcription factor proteins in the regulation of *Arabidopsis* flowering time. Plant Cell 16: 2601–2613.

Oh S, Zhang H, Ludwig P, van Nocker S (2004) A mechanism related to the yeast transcriptional regulator Paf1c is required for expression of the *Arabidopsis* FLC/MAF MADS box gene family. Plant Cell 16: 2940–2953.

Pandey RR, Mondal T, Mohammad F, Enroth S, Redrup L, Komorowski J, Nagano T, Mancini-Dinardo D, Kanduri C (2008) Kcnq1ot1 antisense noncoding RNA mediates lineage-specific transcriptional silencing through chromatin-level regulation. Mol. Cell 32: 232–246.

Pin PA, Benlloch R, Bonnet D, Wremerth-Weich E, Kraft T, Gielen JJ, Nilsson O (2010) An antagonistic pair of FT homologs mediates the control of flowering time in sugar beet. Science 330: 1397–1400.

Quesada V, Macknight R, Dean C, Simpson GG (2003) Autoregulation of FCA pre-mRNA processing controls *Arabidopsis* flowering time. EMBO. J. 22: 3142–3152.

Reeves PH, Murtas G, Dash S, Coupland G (2002) Early in short days 4 a mutation in *Arabidopsis* that causes early flowering and reduces the mRNA abundance of the floral repressor FLC. Development 129: 5349–5361.

Reeves PA, He Y, Schmitz RJ, Amasino RM, Panella LW, Richards CM (2007) Evolutionary conservation of the FLOWERING LOCUS C-mediated vernalization response: Evidence from the sugar beet (*Beta vulgaris*). Genetics 176: 295–307.

Saleh A, Alvarez-Venegas R, Yilmaz M, Le O, Hou G, Sadder M, Al-Abdallat A, Xia Y, Lu G, Ladunga I, Avramova Z (2008) The highly similar *Arabidopsis* homologs of trithorax ATX1 and ATX2 encode proteins with divergent biochemical functions. Plant Cell 20: 568–579.

Schmitz RJ, Hong L, Michaels S, Amasino RM (2005) FRIGIDA-ESSENTIAL 1 interacts genetically with FRIGIDA and FRIGIDA-LIKE 1 to promote the winter-annual habit of *Arabidopsis thaliana*. Development 132: 5471–5478.

Schomburg FM, Patton DA, Meinke DW, Amasino RM (2001) FPA, a gene involved in floral induction in *Arabidopsis*, encodes a protein containing RNA-recognition motifs. Plant Cell 13: 1427–1436.

Schwartz YB, Pirrotta V (2007) Polycomb silencing mechanisms and the management of genomic programmes. Nat. Rev. Genet. 8: 9–22.

Searle I, He Y, Turck F, Vincent C, Fornara F, Krober S, Amasino RA, Coupland G (2006) The transcription factor FLC confers a flowering response to vernalization by repressing meristem competence and systemic signaling in *Arabidopsis*. Genes. Dev. 20: 898–912.

Sheldon CC, Burn JE, Perez PP, Metzger J, Edwards JA, Peacock WJ, Dennis ES (1999) The FLF MADS box gene: A repressor of flowering in *Arabidopsis* regulated by vernalization and methylation. Plant Cell 11: 445–458.

Simpson GG, Dijkwel PP, Quesada V, Henderson I, Dean C (2003) FY is an RNA 3′ end-processing factor that interacts with FCA to control the *Arabidopsis* floral transition. Cell 113: 777–787.

Spitale RC, Tsai MC, Chang HY (2011) RNA templating the epigenome: Long noncoding RNAs as molecular scaffolds. Epigenetics 6: 539–543.

Sung S, Amasino RM (2004) Vernalization in *Arabidopsis thaliana* is mediated by the PHD finger protein VIN3. Nature 427: 159–164.

Sung S, He Y, Eshoo TW, Tamada Y, Johnson L, Nakahigashi K, Goto K, Jacobsen SE, Amasino RM (2006a) Epigenetic maintenance of the vernalized state in *Arabidopsis thaliana* requires LIKE HETEROCHROMATIN PROTEIN 1. Nat. Genet. 38: 706–710.

Sung S, Schmitz RJ, Amasino RM (2006b) A PHD finger protein involved in both the vernalization and photoperiod pathways in *Arabidopsis*. Genes. Dev. 20: 3244–3248.

Swiezewski S, Liu F, Magusin A, Dean C (2009) Cold-induced silencing by long antisense transcripts of an *Arabidopsis* Polycomb target. Nature 462: 799–802.

Tamada Y, Yun JY, Woo SC, Amasino RM (2009) *ARABIDOPSIS* TRITHORAX-RELATED7 is required for methylation of lysine 4 of histone H3 and for transcriptional activation of FLOWERING LOCUS C. Plant Cell 21: 3257–3269.

Trevaskis B, Hemming MN, Dennis ES, Peacock WJ (2007) The molecular basis of vernalization-induced flowering in cereals. Trends Plant Sci. 12: 352–357.

Tsai MC, Manor O, Wan Y, Mosammaparast N, Wang JK, Lan F, Shi Y, Segal E, Chang HY (2010) Long noncoding RNA as modular scaffold of histone modification complexes. Science 329: 689–693.

Wang R, Farrona S, Vincent C, Joecker A, Schoof H, Turck F, Alonso-Blanco C, Coupland G, Albani MC (2009) PEP1 regulates perennial flowering in *Arabis alpina*. Nature 459: 423–427.

Wood CC, Robertson M, Tanner G, Peacock WJ, Dennis ES, Helliwell CA (2006) The *Arabidopsis thaliana* vernalization response requires a polycomb-like protein complex that also includes VERNALIZATION INSENSITIVE 3. Proc. Natl. Acad. Sci. USA 103: 14631–14636.

Xu L, Zhao Z, Dong A, Soubigou-Taconnat L, Renou JP, Steinmetz A, Shen WH (2008) Di- and tri- but not monomethylation on histone H3 lysine 36 marks active transcription of genes involved in flowering time regulation and other processes in *Arabidopsis thaliana*. Mol. Cell Biol. 28: 1348–1360.

Yan L, Loukoianov A, Tranquilli G, Helguera M, Fahima T, Dubcovsky J (2003) Positional cloning of the wheat vernalization gene VRN1. Proc. Natl. Acad. Sci. USA 100: 6263–6268.

Yan L, Loukoianov A, Blechl A, Tranquilli G, Ramakrishna W, SanMiguel P, Bennetzen JL, Echenique V, Dubcovsky J (2004) The wheat VRN2 gene is a flowering repressor down-regulated by vernalization. Science 303: 1640–1644.

Yan L, Fu D, Li C, Blechl A, Tranquilli G, Bonafede M, Sanchez A, Valarik M, Yasuda S, Dubcovsky J (2006) The wheat and barley vernalization gene VRN3 is an orthologue of FT. Proc. Natl. Acad. Sci. USA 103: 19581–19586.

Yun JY, Tamada Y, Kang YE, Amasino RM (2012) ARABIDOPSIS TRITHORAX-RELATED3/SET DOMAIN GROUP2 is required for the winter-annual habit of *Arabidopsis thaliana*. Plant Cell Physiol. 53: 834–846.

Zhang H, van Nocker S (2002) The VERNALIZATION INDEPENDENCE 4 gene encodes a novel regulator of FLOWERING LOCUS C. Plant J. 31: 663–673.

Zhao Z, Yu Y, Meyer D, Wu C, Shen WH (2005) Prevention of early flowering by expression of FLOWERING LOCUS C requires methylation of histone H3 K36. Nat. Cell Biol. 7: 1256–1260.

Zhao J, Sun BK, Erwin JA, Song JJ, Lee JT (2008) Polycomb proteins targeted by a short repeat RNA to the mouse X chromosome. Science 322: 750–756.

5 Temperature and light signal integration

Harriet G. McWatters, Gabriela Toledo-Ortiz, and Karen J. Halliday

5.1 Introduction

Plants are sessile, photosynthetic organisms. They must therefore cope with changes in their local environment by adjusting their physiology appropriately – unlike animals they cannot avoid adverse conditions by moving to another location. Perhaps as a consequence, their growth and development is extremely plastic, and they have developed sensitive sensory mechanisms to constantly monitor their surroundings to respond to changes in ambient conditions, in particular to variation in light and temperature. Plants are sensitive to a range of wavelengths of light and to subtle changes in temperature, and their daily and seasonal cycles of photosynthesis and carbohydrate usage, growth, and reproduction are precisely aligned to their environment so as to make best use of resources and increase fitness (Dodd et al. 2005).

In the laboratory, plants are grown under carefully controlled conditions with precise regulation of light intensity and photoperiod, temperature, and humidity. However, under natural diurnal conditions, major environmental factors such as temperature, light, and humidity change constantly. The rotation of the planet produces the predictable changes of day and night. Photoperiod is linked to latitude and season as the planet tilts on its axis in its yearly orbit around the sun. Photoperiod is thus an unvarying time signal for plants to coordinate their life history with the yearly seasonal changes. Temperature changes are associated with the daily and seasonal variations in photoperiods, as days tend to be warmer than nights and summers than winters. However, this is a general trend, and on any given day, plants may experience a wide combination of light levels and temperatures. Figure 5.1 shows ground temperature, solar flux, and relative humidity throughout August 2011 in Edinburgh, UK (latitude 55.921568, longitude 3.174394, altitude ~105 m). Absolute level of solar flux changes rapidly and on multiple timescales: rapid, as clouds pass in front of the sun, and longer term across the daylight hours and reflecting time of day.

Temperature and Plant Development, First Edition. Edited by Keara A. Franklin and Philip A. Wigge.

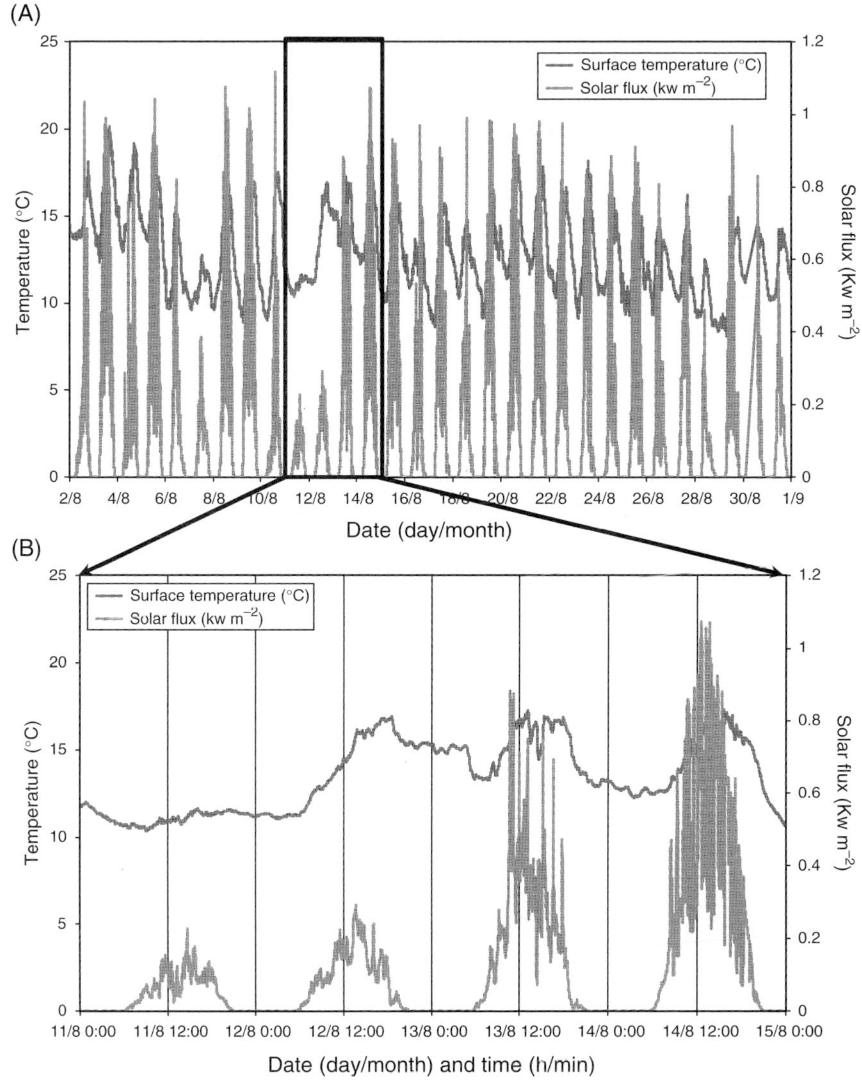

Figure 5.1 Variation in light and temperature provides a constantly changing environment for plants. There is an underlying pattern reflecting day and night, but a high degree of noise is superimposed over this; thus, a plant will have to be constantly monitoring and responding to environmental change. (A) Ground temperature and solar flux across the month of August 2011 in Edinburgh, UK. (B) Effect of rainy, cloudy days. Expanded panel showing the extent of day on day variation across consecutive days. Note the very small temperature range on the 11th of August and the rapid changes in solar flux on the 14th of August. (C) Relative humidity across August 2011. Weather data courtesy of Massimo Vieno, School of GeoSciences, Edinburgh University. For color detail, please see color plate section.

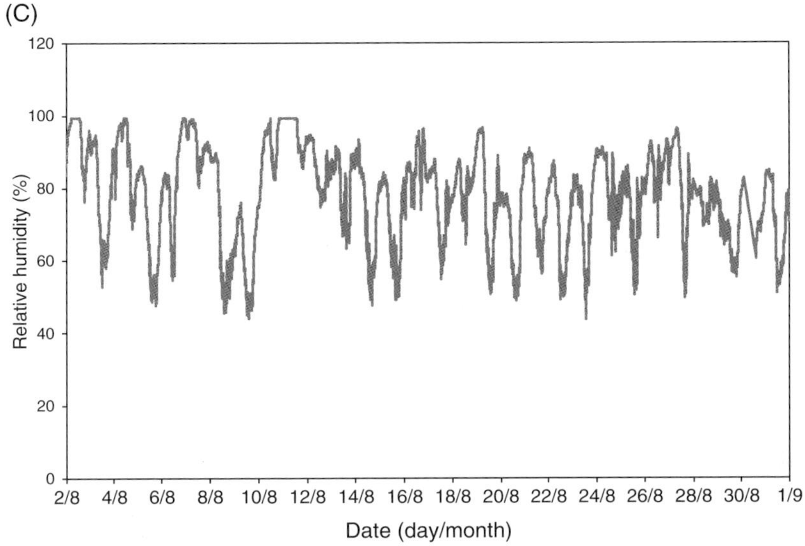

Figure 5.1 (*Continued*)

But nonetheless, there is a clear 24-h pattern to the data apparent even on the overcast days of the 11th and 12th of August (Figure 5.1A, B). Although temperature in large part tracks solar flux, the pattern is more apparent on bright days than on overcast ones. Indeed, the difference between daytime and nighttime temperatures ranges from 1°C to 10°C (Figure 5.1A, B). The pattern of relative humidity also shows a high degree of noise superimposed upon the 24-h pattern (Figure 5.1C).

The weather data in Figure 5.1 make the point that temperature and weather are unpredictable and variable on all timescales. Especially at high latitudes, winters may strike quickly with wide variation in onset date (and hence photoperiod) between years. Successful plant growth and reproduction thus requires integration of light and temperature information on a daily and seasonal basis. For instance, temperature, at the level of the plant, can vary within minutes by up to 20°C due to the effects of sun and wind (McClung and Davis 2010), and *Arabidopsis* accessions across Western Europe can tolerate temperature gradients of more than 30°C (Wilczek et al. 2009). Plants have developed the ability to monitor photoperiods, ambient light levels, and temperature and to respond rapidly to changes by enabling major changes in physiology that provide thermotolerance and cold acclimation. In so doing, different species of plants are able to live across most areas of the Earth's surface, including those experiencing some of the widest ranges of temperature between night and day, as in the North African deserts, or from summer to winter, as in the coastal fringes of Antarctica.

Within the optimal range of temperatures for development, plants use mechanisms such as the circadian clock to buffer the effects of variation across the day and prevent inappropriate responses to random change. However, temporal trends in temperature, such as warming in spring, are also detected and used to coordinate major aspects of plant physiology with the seasons and time major life-history events, such as budburst in deciduous trees or the transition from vegetative growth to flowering in small annuals like *Arabidopsis*. Temperature also influences metabolic processes (i.e., photosynthesis and respiration), plant growth, and biomass production and allocation, which are promoted by high temperature and restricted in the cold, for example (Blazquez et al. 2003; Halliday et al. 2003).

All these events require coordination of the interplay between the environmental temperature signals and the endogenous transcriptional networks, hormonal pathways, and other environmental perception pathways of the plant. Therefore, the processes controlled by temperature cannot be studied in isolation. In particular, current research indicates that in nature, environmental inputs from light and temperature are greatly correlated with each other and with daily timing (Mas and Yanovsky 2009).

Growth is not a simple event but the result of the interplay of many developmental pathways with the environment. *Arabidopsis* plants under a long-term exposure to low temperatures (17°C) are smaller, with a decreased shoot development, thicker leaves, and less leaf area per unit mass, compared to warmer-grown plants (22°C) (Storkey 2004; Atkin et al. 2006) (Figure 5.2). On the other hand, at high temperatures (27–28°C), plants show stem or petiole elongation, hyponasty, accelerated flowering, and reduced biomass (Patel and Franklin 2009). These warm-temperature phenotypes are similar to ones observed in response to vegetational shade or genetic depletion of phytochromes (Franklin 2008). Figure 5.2 illustrates this phenomenon in an *Arabidopsis* mutant lacking phytochromes A, B, and D (*phyABD*). These morphological changes have been shown to involve the circadian clock, which controls the timing of cell elongation, as well as fluctuations of the levels of several hormones (including auxin, brassinosteroids, and gibberellic acid (GA)), and modulation of the activity of various classes of transcriptional factors and transcriptional co-regulators (i.e., Phytochrome-Interacting Factors (PIFs), Elongated Hypocotyl 5, DELLA proteins) (Patel and Franklin 2009).

Moreover, the plant response to temperature is intimately involved with human activity. The average first flowering time of 385 British plants advanced by 4.5 days between the 1950s and the 1990s, with most of that change occurring during the 1980s (Fitter and Fitter 2002). The advance was strongly associated with concomitant warming of the English climate (UK Meteorological Office data cited by Fitter and Fitter 2002), underlining the effect of warmer spring times upon plant communities. It has, for

Figure 5.2 Phenotype of 46-day-old WT (Ler) on left and *phyABD* mutant *Arabidopsis* plants on right grown in white light, under short days (8 h light:16 h darkness) at 21°C. For color detail, please see color plate section.

example, been suggested that if current projections of climate change are correct, up to a third of European plant communities will face extinction by 2080 (Thuiller et al. 2005). Such observations underline the importance of better understanding how plants sense and respond to temperature, both as 'pure' and 'applied' plant science, given the current trends in global warming and the challenge this poses to agriculture and food security (Battisti and Naylor 2009).

5.2 Convergence points for light and temperature sensing

Many of the temperature-regulated developmental processes have a link with light signaling, indicating that these complex pathways must converge to allow successful plant development. Recent work, mostly using the model plant *Arabidopsis thaliana*, has started to delve into the molecular genetic pathways governing transcriptional regulation and signal transduction in response to changes in light and temperature.

Plant photomorphogenesis is known to involve at least three main classes of photoreceptors: the red (R)- and far-red (FR)-perceiving phytochromes (five phytochromes in *Arabidopsis*, phyA through phyE), the UV-A/blue light-absorbing cryptochromes (cry1–cry3), and the phototropins (phot1, phot2) (Nagy and Schafer 2002). Phytochrome and cryptochrome-regulated responses are strongly influenced by the ambient temperature, but the capacity of these photoreceptors to remain active across a temperature range suggests a convergence of light and temperature signaling at the molecular level (Franklin et al. 2003; Halliday and Whitelam 2003; Halliday et al. 2003; Penfield et al. 2005; Heschel et al. 2007). Among the molecular components identified to date, the PIFs, the DELLA proteins, and ELONGATED HYPOCOTYL 5 (HY5) are emerging as convergence points for light and temperature signals.

5.3 Phytochrome-Interacting Factors as signal integrators

The PIFs are a small family of closely related basic helix–loop–helix (bHLH)-type transcription factors that act to repress germination and promote skotomorphogenesis and shade avoidance. As their name suggests, they act immediately downstream of the phytochromes. Light-activated (Pfr form) phytochrome moves from the cytosol into the nucleus where it interacts with a variety of targets including the PIFs. PIFs are capable of binding the active Pfr forms of phytochrome B (phyB) via a conserved N-terminal sequence, the Active Phytochrome B-binding (APB) motif (Khanna et al. 2004), and PIF1 and PIF3 can also bind to phyA (Huq et al. 2004; Al-Sady et al. 2008). PIFs recognize a core DNA G-box motif in the promoters of their target genes (Leivar and Quail 2011). Together the PIFs regulate expression of over one thousand genes, mostly as positive activators of expression, although they may also act as transcriptional repressors (Castillon et al. 2007).

PIFs accumulate in dark-grown seedlings. In the light, their levels drop rapidly because binding by active phytochrome results in their phosphorylation, ubiquitination, and degradation via the ubiquitin–proteasome system (Park et al. 2004; Al-Sady et al. 2006; Shen et al. 2008). This may be the primary event in phytochrome signaling, with degradation of PIFs being the means by which phytochromes trigger the transition from skotomorphogenic to photomorphogenic growth. Consistent with this hypothesis, a quadruple *pif1pif3pif4pif5 (pifQ)* mutant shows constitutively photomorphogenic growth in darkness (Leivar et al. 2008b). Both PIF4 and PIF5 are integral in regulating shade avoidance (Franklin 2008; Lorrain et al. 2009), and there is a considerable overlap between genes regulated by PIF4 and genes known to regulate growth and shade avoidance

(Nozue et al. 2011). In R light, PIF4 and PIF5 are selectively degraded due to interaction with activated phyB. In contrast under shade-mimicking conditions, their levels remain high, and thus they are able to continue promoting growth (Nozue et al. 2007; Lorrain et al. 2008). The effects of PIF4 and PIF5 are opposed by LONG HYPOCOTYL IN FAR-RED1 (HFR1) which acts to inhibit growth (Hornitschek et al. 2009). HFR1 is a PIF-like (PIL) transcription factor regulated by cry1 and phyA photoreceptors acting principally in the perception of blue and FR light, respectively (Fairchild et al. 2000; Fankhauser and Chory 2000; Duek and Fankhauser 2003; Zhang et al. 2008). In shade or FR-enriched light, the growth-promoting effects of PIF4 and PIF5 are antagonized by HFR1, which inhibits action of the transcription factors by forming non-DNA-binding heterodimers with the PIFs (Hornitschek et al. 2009; Foreman et al. 2011) and thus holds the shade-avoidance response in balance.

PIFs play a far wider role than merely regulating plant responses to light. Their extended functions include controlling seed germination, plant architecture, regulation of plant metabolism, and cell fate. By being functionally diverse, they are emerging as transcriptional integrators of information from multiple signaling pathways to control plant morphogenesis (Leivar and Quail 2011). In particular, their role in regulating the intersection between light and hormone (gibberellic acid (GA), auxin) signaling is probably one of their key roles in controlling growth (Franklin et al. 2011; Nozue et al. 2011; Hornitschek et al. 2012; Leivar et al. 2012; Li et al. 2012; Sellaro et al. 2012; Sun et al. 2012a). In the case of GA signaling, PIF3 and PIF4 are known targets of the DELLA proteins, GAI and RGA. DELLAS are nuclear GA-regulated proteins that accumulate in the absence of GAs blocking the transcriptional response to the hormone (de Lucas et al. 2008; Feng et al. 2008). The discovery that the auxin-responsive gene IAA29 is a direct target of PIF4 (Koini et al. 2009) implicated this member of the family in the auxin signaling network. Furthermore, PIF4 was identified as a gene upregulated by a short-term increase in temperature (Stavang et al. 2009). The evidence suggests that PIF4, in particular, is a central integration point for various light and temperature signaling pathways and thus connects the environmental inputs which trigger and the hormonal responses which produce growth.

Exposing *Arabidopsis* seedlings to high temperature (27–29°C) produces a noticeable phenotype, elongated hypocotyls, extension of axes, leaf hyponasty, and early flowering, which is similar to the classic shade-avoidance response and reminiscent of the appearance of phytochrome mutants (Casal 2012). Over a decade ago this response was shown to require auxin. Auxin response or transport mutants failed to change growth patterns in response to high temperature, and an increase in ambient temperature promoted accumulation of IAA (Gray et al. 1998). Recent work, summarized in the following text, has detailed the role of PIFs, especially PIF4, in controlling growth at high temperatures by modulating the level of auxin. Franklin et al. (2011) showed that PIF4 regulates auxin levels by controlling the expression of auxin biosynthesis genes and SMALL AUXIN UP RNA (SAUR) genes, which promote elongation at high temperature. In this way, PIF4 provides one of the molecular links between the phenotype of elongated growth exhibited by *Arabidopsis* at high temperature and auxin hormone levels/signaling components.

Another control point of auxin signaling modulated by both PIF4 and PIF5 at high temperature is the upregulation of the auxin biosynthetic gene *YUCCA8* (Nozue et al. 2011; Sun et al. 2012b). Transfer of wild-type (WT) plants to elevated temperature was shown to cause a transient increase expression of *PIF4* and of *YUC8* and a marked rise in the level of auxin. Overexpression of *PIF4* in plants grown at ambient temperature produced increased activity of *YUC8*, high levels of endogenous auxin, as well as the classic long hypocotyl phenotype associated with excess auxin (Sun et al. 2012b). Conversely, the *pif4pif5* double mutant does not show an elongated hypocotyl at high growth temperature and is less sensitive to auxin (Nozue et al. 2011). *In vitro*, PIF4 binds directly to the G-box in the *YUC8* promoter and activates its expression. The *yuc8* mutation suppresses the long hypocotyl phenotype of *pif4* and also reduces hypocotyl elongation in response to high temperature (Sun et al. 2012b). *shy2-2*, which produces a stabilized form of the SHY2/IAA3 protein, also suppresses hypocotyl elongation in *pif4*, but other gain-of-function auxin mutants did not, indicating selective action in the PIF4–auxin signaling pathway regulating growth at high temperature (Sun et al. 2012b).

PIF4 is modified posttranslationally by phosphorylation. Although at moderate temperature (15°C) phosphorylated PIF4 is rapidly degraded in light (Lorrain et al. 2008; Foreman et al. 2011), high temperature (25°C) promotes the accumulation of phosphorylated PIF4, even in the presence of light (Foreman et al. 2011). This implies that the different activities of PIF4 in response to light and temperature changes could be mediated by phosphorylation. Moreover, condition-dependent phosphorylation could have opposing effects on protein stability. Further analysis is required to determine whether the same residues are targeted for phosphorylation in different light and temperature conditions.

Light receptor pathways are known to be temperature regulated and to buffer the plant against the effects of temperature changes. At the molecular level, the HFR1–PIF system provides a good example of how buffering could be achieved. PIF4, and its close relative, PIF5, forms heterodimers with HFR1 (Sessa et al. 2005; Hornitschek et al. 2009) which inhibits their DNA-binding activity. Mutant seedlings deficient in HFR1 showed exceptionally elongated hypocotyls under blue light at 25°C but not at 15°C, indicating joint regulation of light and temperature signaling (Foreman et al. 2011). Under warm temperature and blue light, phosphorylated PIF4 accumulates (Foreman et al. 2011). In these conditions, HFR1 stabilization by cryptochrome leads to a strong suppression of PIF4 activity, through dimerization. As phyB also promotes PIF4 degradation across temperatures, dual action from the light receptors directly (phyB) or via activation of HFR1 (crys) strongly antagonizes PIF4 activity.

In the warm, although both HFR1 and PIF4 levels increase, the repressive effect of HFR1 on PIF4 gets stronger as the temperature rises. Likewise, temperature and phyB have antagonistic effects over PIF4. Thus, blue light signaling via crys and HFR1 and R light signaling via phyB are both modulated by temperature and converge at PIF4. This ability to buffer temperature-induced changes in PIF4 activity is important as *phy* and *cry* mutants are compromised with respect to biomass and viability (Foreman et al. 2011). Growth inhibition in response to temperature is a complicated response regulated in parallel by several pathways. The involvement of auxin has been discussed previously. The phytohormone GA also acts to regulate growth, and light targets GA synthesis to repress hypocotyl elongation and photomorphogenesis (Achard et al. 2007). Growth responses to GA are regulated by the DELLA proteins. DELLAs restrain plant growth, whereas GA promotes growth by targeting the DELLAs for destruction (Harberd et al. 1998; Harberd 2003; Achard et al. 2008). Double mutants deficient in the DELLA genes *GA1* and *RGA1* displayed an elongated hypocotyl phenotype at high temperatures, and even a pentuple DELLA mutant retained some temperature sensitivity, indicating that the auxin and brassinosteroid pathways act independently of DELLAs (Stavang et al. 2009). Expression of genes in the GA biosynthesis pathway increased following transfer to high temperature, indicating increased activity of this pathway (Stavang et al. 2009). DELLA proteins oppose PIF4 function by binding to the DNA-recognition domain and hence blocking transcriptional activity (de Lucas et al. 2008). As GA causes destabilization of DELLAs, high levels of GA lead to an increase in PIF4 activity and promote growth.

5.4 ELONGATED HYPOCOTYL 5 (HY5): A cool operator

HY5 is one of the central modulators of gene expression for coordination of light signals, hormone signaling, control of plant development, and plant resistance to osmotic and ionic stress (Chen and Xiong 2008). As a bZIP transcription factor, HY5 is involved in transducing multiple photoreceptor signals to promoters and activating photomorphogenesis (Lau and Deng 2010). Chromatin-immunoprecipitation (ChIP)–chip experiments have shown that HY5 binds to the promoters of up to 9000 genes and significantly affects the expression of at least 12% of its targets (Lee et al. 2007; Catala et al. 2011). To regulate plant growth, HY5 is a point of convergence for many signaling pathways, including hormone biosynthesis. In particular, HY5 controls the levels of ethylene (Li et al. 2011) and acts in the antagonistic ABA pathway to induce ABA-dependent gene expression during germination (Chen and Xiong 2008).

A transcription factor protein microarray identified 20 potential partners for HY5 (Gong et al. 2008), indicating that in some cases HY5 exerts and diversifies its effects by partnership with other transcription factors. Among the identified partners are CIRCADIAN CLOCK-ASSOCIATED 1 (CCA1) for the circadian control of *CHLOROPHYLL A-/B-BINDING PROTEIN* (*CAB*) genes (Andronis et al. 2008) and FAR-RED ELONGATED HYPOCOTYL 3 (FHY3)/FAR-RED IMPAIRED RESPONSE 1 (FAR1) for the regulation of phyA signaling (Li et al. 2010). HY5 principally acts during the day to activate light-responsive genes, and mutations at the *HY5* locus result in a failure to inhibit hypocotyl elongation even in light (Ang and Deng 1994; Lau and Deng 2010).

Posttranslationally, HY5 is degraded in the nucleus in darkness by the E3 ubiquitin ligase CONSTITUTIVE PHOTOMORPHOGENIC 1 (COP1) (Osterlund et al. 2000). The action of COP1 on HY5 is positively regulated by the repressor of photomorphogenesis, SUPPRESSOR OF PHYA 1 (SPA1) (Saijo et al. 2003). In blue light, the photoreceptor CRY1 interacts with SPA1 to suppress the SPA1–COP1 interaction and suppresses COP1 activity, leading to an increase in HY5 levels (Liu et al. 2011). Expression of *HY5* is also positively regulated by phyA (Tepperman et al. 2001), and HY5 is an important component in the feedback regulation of phyA signaling (Li et al. 2010).

Besides its key role in promoting photomorphogenesis, HY5 is central in protecting plants against abiotic stress such as UV-B and cold (Favory et al. 2009; Gruber et al. 2010). Transcript profiling has indicated that HY5 regulates about 10% of all cold-inducible genes, including many not inducible by light, showing a role for HY5 in cold acclimation independent of light signaling pathways (Catala et al. 2011). Among the cold-induced targets of HY5 is *CAB1* (Capel et al. 1998; Catala et al. 2011). HY5 induces *CAB1* expression by binding a low-temperature-responsive element (LTRE) in the Z-box (Catala et al. 2011). Also, in response to cold stress, HY5 induces anthocyanin production (Zhang et al. 2011c) to protect the plant against damage caused by the accumulation of reactive oxygen species (ROS). In this context, *HY5* accumulates rapidly in the cold through an abscisic acid (ABA)- and C-REPEAT BINDING FACTOR (CBF)-independent pathway and remains stable even in the dark at low temperature via a mechanism that involves increasing the pool of HY5 protein by exclusion of COP1 from the nucleus (Catala et al. 2011).

The switch between photomorphogenic and skotomorphogenic growth is triggered by light, which activates HY5 but causes degradation of PIFs. GA, however, counters the effect of light by negatively regulating HY5 levels and also by stabilizing PIFs by reducing the negative effects of DELLA proteins (Feng et al. 2008; de Lucas et al. 2008; Gallego-Bartolome et al. 2011). In fact, half the targets of the DELLA protein GAI are also regulated by HY5 or by PIFs or by both (de Lucas et al. 2008). At low temperature, DELLA proteins accumulate, in part through activation of the CBF/DREB1 pathways (see Section 5.7). DELLAs contribute to freezing tolerance but also inhibit growth, acting in opposition to the growth-promoting effects of GA (Harberd 2003). DELLA degradation is stimulated by GA, and thus the rate of degradation is dependent upon GA level. Low temperature causes a reduction in GA level due to an increase in GA 2-oxidases, the enzymes that inactivate GA (Achard et al. 2008), and the resulting increase in DELLA proteins provides the primary inhibition of growth in the cold. This pathway involves HY5 and its homologue, HOMOLOGUE OF HY5 (HYH), which upregulate *GA2ox1* (Zhang et al. 2011a). Thus, the accumulation of DELLA proteins at low temperature also requires *HY5*.

The interplay between GA and HY5 is further illustrated by the effect of blocking GA biosynthesis in dark-grown seedlings. Normally this causes deetiolation and the appearance of photomorphogenic growth. However, *hy5* seedlings are resistant to this block and show reduced unfolding of cotyledons, higher levels of *CAB2* transcript abundance, and hypocotyl elongation inhibition, relative to WT plants (Alabadi and Blazquez 2008). Because stability of HY5 is enhanced when GA levels fall, as they do in the dark or in the cold, plants can mount an appropriate growth pattern in response to environmental change.

5.5 Light and temperature converge at the circadian oscillator

The circadian clock acts as a reference that allows plants to measure changes in the timing of external events. The function of the circadian clock is to generate physiological rhythms closely synchronized with the 24-h period of the environment. The biological rhythms are generated by oscillation of molecular components whose transcription or activities are rhythmically controlled and interlocked by transcriptional–translational loops (Salome and McClung 2005). Circadian clock architecture and temperature interactions are discussed in detail in Chapter 6.

The primary time signal (zeitgeber) for time of day or year is light. Plants, like most other organisms, possess a circadian clock that is entrained by the pattern of light and dark. Light and temperature signals interact with the circadian clock to regulate rhythmicity of physiological and developmental processes and anticipate environmental changes. The circadian oscillator of *Arabidopsis* is a complicated affair made up of several interlocked feedback loops, each of which is phased either to the morning (i.e., to sunrise) or to the evening (i.e., to sunset). A full description of the clock is beyond the scope of this review but may be found in Pokhilko et al. (2012) or in recent reviews (e.g., Harmer 2009; McWatters and Devlin 2011). However, in brief, the core clock components include two highly homologous MYB transcription factors LHY and CCA1, which are expressed in the morning and which are negatively regulated by the five members of the PSEUDO-RESPONSE REGULATOR (PRR) family, *TOC1* (also known as *PRR1*), *PRR3*, *PRR5*, *PRR7*, and *PRR9*. CCA1 and LHY bind to the evening element (EE), a conserved element in the promoters of target genes, both within the clock and downstream from it (Harmer et al. 2000). *ELF3*, *ELF4*, and *LUX* are important regulators of the clock gene expression during the night (Onai and Ishiura 2005; Kolmos et al. 2009; Dixon et al. 2011; Helfer et al. 2011). ELF3, ELF4, and LUX proteins form a complex called the evening complex (EC) which binds to the promoters of target genes (Nusinow et al. 2011) (such as *PRR9* and *LUX*), suppressing their expression (Dixon et al. 2011; Helfer et al. 2011).

Many, perhaps most, physiological pathways are influenced by the circadian clock; up to 89% of *Arabidopsis* genes show rhythmic expression under one set of environmental conditions, some being driven by light or temperature cycles but others by the endogenous circadian clock itself. It is the latter which persist in constant conditions (Harmer 2009). In the absence of daily photoperiods, plants can entrain to temperature cycles (thermocycles) with as little as 4°C difference between warm 'days' and cold 'nights' (Somers et al. 1998). The warm period is considered the 'day' and the cold period the 'night' because, under natural conditions, days are often warmer than nights and so light and temperature cues reinforce each other. In temperature cycles, 50% of *Arabidopsis* genes showed rhythmic expression in constant light, whereas 45% were rhythmic in light/dark cycles at constant temperature (Michael et al. 2008), indicating the stimuli are of similar strengths. Light input to the circadian clock is mainly via the phytochrome and cryptochrome photoreceptors, and mutations in these lead to alterations in circadian phenotypes. It is not clear whether there is a dedicated receptor for temperature input to the plant clock. However, entrainment to temperature cycles requires ELF3 as well as members of the PRR family (McWatters et al. 2000; Thines and Harmon 2010). ELF3 acts in both light and temperature pathways to ensure rhythmicity of input even under constant conditions. On the other hand, the observation that *prr7* and *prr9* mutants do not reset their clocks following a cold pulse suggests the involvement of PRRs in circadian temperature responses (Salome and McClung 2005).

The rate of a biochemical reaction usually doubles with a 10°C increase in temperature – if the clock doubled in speed under such a circumstance, it would be unfit for purpose. Thus, in order to keep to the correct time, the circadian clock must be buffered against temperature change. This phenomenon is called 'temperature compensation' and is one of the canonical properties of a circadian clock (Pittendrigh 1954). Although the free-running period of the *Arabidopsis* circadian clock does shorten slightly as temperature rises, it is relatively unchanged across the temperature range 12–27°C (Edwards et al. 2005, 2006).

The identification of molecular components involved in temperature compensation of the plant clock is an active field of research. A noticeable feature of many of the genes involved in temperature compensation of the plant clock is that they are light regulated (McClung and Davis 2010). The free-running period of the clock is closely tied to the strength of light input. Aschoff's rule (Aschoff 1979) states that the free-running period of a day-active organism decreases as light intensity increases. The concept of 'photo-compensation' has not been much considered in circadian biology, but, formally, a clock needs to be buffered against changes in light intensity to keep correct time, and it is clear that the system is indeed buffered against changes in light intensity as when *Arabidopsis* plants are exposed to R light fluence rates ranging from 1 to 200 μmol m^{-2} s^{-1}, the free-running circadian period only decreases from 27 to 23 h (e.g., Devlin and Kay 2000; Covington et al. 2001; Kim et al. 2008). The daily signals of light and temperature usually coincide (warm days, cold nights), and dual sensitivity of components, such as *CCA1*, provides a means to integrate environmental information (Salome and McClung 2005).

Although a quantitative trait locus analysis identified *FLOWERING LOCUS C* (*FLC*) as a gene required for temperature compensation at high temperature and suggested that *LUX ARRHYTHMO* (*LUX*) was the target of FLC (Edwards et al. 2006), most work on the subject has focused upon genes making up the core clockwork. The first mechanism proposed for temperature compensation in plants, based upon an analysis of free-running periods of clock mutants at different temperatures, involved temperature-dependent differential feedback from GIGANTEA (GI) and ZEITLUPE (ZTL), which both act in the clock in the evening to regulate *TIMING OF CAB1* (*TOC1*), onto the morning clock components LHY and CCA1(Gould et al. 2006).

All the *PRRs* play roles in the circadian clock (Mizuno and Nakamichi 2005; Nakamichi et al. 2005, 2007; Matsushika et al. 2007); however, *PRR9* and *PRR7* are required for a circadian response to temperature cycles since the *prr7prr9* double mutant cannot entrain to temperature (Salome and McClung 2005). This work was extended to implicate these two genes in the temperature-compensation mechanism. Unlike WT plants, the free-running period of the *prr7prr9* double mutant increased with rising temperature, an overcompensation phenotype not seen in other clock mutants. Both PRR7 and PRR9 negatively regulate *CCA1* and *LHY*. In *prr7prr9* plants, levels of *CCA1* and *LHY* expression increased with temperature, with *LHY* showing a strongly linear response. Moreover, targeted knockdown of *LHY* and *CCA1* by microRNA abolished the overcompensation phenotype and caused *prr7prr9* plants to maintain a constant period across the temperature range 12–27°C (Salome et al. 2010).

Other studies of temperature compensation have been centered on posttranslational modification of the highly homologous morning clock components LHY and CCA1. In the fungus *Neurospora crassa*, a longtime model organism for circadian biology, temperature compensation involves phosphorylation of the clock protein FREQUENCY (FRQ) by CK2 which causes preferential degradation of FRQ at high temperatures (Mehra et al. 2009). Recent research on *Arabidopsis* suggests a similar mechanism may be in play in plants. In *Arabidopsis*, CK2 phosphorylates CCA1 and LHY (Sugano et al. 1998) and this is required for the correct functioning of CCA1 in the oscillator mechanism (Sugano et al. 1998; Daniel et al. 2004). Portoles and Mas (2010) showed that CK2 affects the transcriptional activity of CCA1, but does not affect its accumulation nor its subcellular localization. CK2 acts in a temperature-dependent manner to antagonize the binding of CCA1 to its targets by altering its phosphorylation state. The dephosphorylated protein is better able to bind to DNA, but at 27°C the level of phosphorylated CCA1 protein increases, creating a temperature-dependent balance between CCA1-binding activity and CK2 phosphorylation (Portoles and Mas 2010). Therefore, it is suggested that CK2 action on CCA1 has a breaking effect upon the speed of the circadian oscillator, allowing temperature compensation. In this context, although at high temperatures CCA1's binding to its target DNA is enhanced (reducing clock speed), so is the opposing function of CK2 (slowing the clock). Thus, the net effect is the reduction of the efficiency of CCA1 binding at high temperature and maintenance of period length across a range of ambient temperature (Portoles and Mas 2010).

In *Arabidopsis*, recent reports show that alternative splicing is a novel mechanism impacting the activity of clock components in response to temperature. Alternative splicing regulates the diversity of the transcriptome and proteome. In response to changes in temperature, key clock components present dynamic alterations in alternative splicing variants. Among the identified clock genes with temperature-dependent alterative splicing are CCA1 (Seo et al. 2012), LHY, and PRR7 (James et al. 2012). CCA1-dependent cold-acclimation responses depend on the self-regulation of CCA1 activity by a splice variant (CCA1β) (Seo et al. 2012). CCA1β forms nonfunctional heterodimers with reduced DNA-binding affinity and therefore interferes with the activity of CCA1α–CCA1α homodimers, LHY–LHY homodimers, and CCA1α–LHY heterodimers. Low temperatures suppress CCA1 alternative splicing and reduce CCA1β production. Overexpression of CCA1α in transgenic plants renders enhanced freezing tolerance, but 35S::CCA1β lines are sensitive to cold temperatures (Seo et al. 2012).

In a related manner, the alternative spliced variants of LHY and PRR7 contribute to functional changes in clock gene expression in temperature transitions by producing nonfunctional transcripts and/or inducing decay. In the case of LHY, temperature effects on alternative splicing contribute to a decrease in LHY transcript abundance during cooling (James et al. 2012). Since identified components of the clock with temperature-dependent alternative spliced variants are also light regulated, alternative splicing is an additional mechanism that could be involved in the integration of light, temperature, and circadian responses. Further investigation of this possibility may render a novel pathway for integration of light, temperature, and clock signals.

Temperature compensation of the circadian clock also appears to involve at least one other MYB transcription factor, REVEILLE8 (RVE8), which, like RVE1, CCA1, and LHY, binds the EE in promoters (Rawat et al. 2011). Mutants deficient in RVE8 displayed long free-running periods, relative to WT plants, and RVE8-ox plants short periods but only at the higher end of the range of ambient temperature, indicating a disruption of temperature compensation (Rawat et al. 2011). Interestingly, perhaps, RVE8 does not appear to influence clock speed by acting on the light input pathway. Levels of RVE8 were increased in a *prr5prr7prr9* mutant and it is tempting to conclude that RVE8 is part of the same signaling mechanism.

So far we have considered the mechanism(s) by which the plant clock is buffered against moderate changes in ambient temperature. It is important to remember that these studies encompass only part of the temperature range (roughly 10–30°C), below temperatures inducing heat shock responses and above those inducing cold acclimation. Outwith these limits, the clock appears to be much less able to maintain a circadian period. In fact, the plant clock appears to stop at low temperature in deciduous trees such as sweet chestnut (*Castanea sativa*). Cycling of levels of clock components cannot be detected in samples collected from trees during their winter dormancy or chilled to 4°C, although there are robust rhythms of expression at 22°C, or in samples collected during the summertime (Ramos et al. 2005; Ibanez et al. 2008). This pattern of high, constant expression of clock genes in cold conditions was also observed in aspen (*Populus tremula* × *P. tremuloides*) (Ibanez et al. 2010), suggesting it is a general feature of deciduous trees. Furthermore, *Populus* trees in which *PttLHY* or *PttTOC1* expression had been knocked down by RNAi showed changes in the pattern of bud set (the point when growth ceases and the tree enters winter dormancy) and freezing tolerance (Ibanez et al. 2010). Although in each case, clock output (measured by *CCR2::LUC* expression) became arrhythmic, indicating the failure of the clock, disrupting *LHY* expression delayed growth cessation and bud set and rendered trees less tolerant to freezing, whereas knocking down *TOC1* caused the opposite phenotype, improved freezing tolerance and early bud set.

Similar effects have been found in *Arabidopsis* where cycling of clock components at 4°C is strongly dependent upon the presence of a light/dark cycle. Under constant light, even core clock genes such as *CCA1*, *LHY*, *GI*, and *PRR5*, *PRR7*, and PRR*9* become arrhythmic at low temperature (Bieniawska et al. 2008; Espinoza et al. 2008). It is not just clock transcription that becomes arrhythmic at low temperature. Clock-controlled metabolites too show major changes in synthesis and degradation, implying that the major metabolic reconfiguration at chilling temperatures is in large part a result of altered clock function (Espinoza et al. 2010). It is unclear whether the changes in transcript and metabolite levels resulting from the strong damping of the circadian clock affect plant growth. However, this is not unlikely as the clock makes a strong contribution to control of growth under other conditions (Ni et al. 1999; Dodd et al. 2005) through regulation of pathways promoting chlorophyll biosynthesis (Ni et al. 1999) and starch degradation (Gibon et al. 2004; Blasing et al. 2005). What is certain is that clock mutations alter plant resistance to cold: for example, *cca1* and *cca1lhy* mutants of *Arabidopsis* showed reduced freezing tolerance and cold acclimation (Espinoza et al. 2010; Dong et al. 2011).

In autumn, deciduous trees at temperate latitudes enter a period of winter dormancy. They cease growth, shed their leaves, and undergo bud set and cold acclimation. Many annual plants also overwinter, including the many ecotypes of *Arabidopsis* which germinate in the late summer and autumn and overwinter as rosette plants to flower early the following spring. Such plants need to adjust their physiology to the challenges of winter: low temperature and reduced light. The primary cue for this change is the shortening photoperiod, as the pattern of photoperiodic change is predictable at a given latitude. Although freezing temperatures provide the greatest physical challenge to plants during the winter, temperature change is a much less accurate signal of seasonal change as the precise pattern of temperature variation is erratic and varies year on year. By relying on photoperiodic cues, plants are able to anticipate and prepare for seasonal environmental changes.

Overexpression of phyA in hybrid aspen alters the critical day length at which trees enter winter dormancy, suggesting that this photoreceptor controls perception of photoperiod (Olsen and Junttila 2002; Karlberg et al. 2011). Cold acclimation was also prevented by overexpression of phyA.

More recently, an increase in critical day length and early onset of growth cessation have been observed in aspen trees in which expression of *PttPHYA* is inhibited by antisense RNA (Kozarewa et al. 2010). The mechanism by which phyA regulates growth cessation and bud set in aspen involves an alteration in the timing of *CONSTANS* (CO) expression with subsequent effects on *PttFKF1* and *PttFT* (Kozarewa et al. 2010). Therefore, bud set and growth cessation occur in the absence of the coincidence of a peak in CO expression during the light period. This implies that an external coincidence mechanism (Burning 1936; Pittendrigh and Minis 1964) regulates growth cessation and bud set in deciduous trees and, moreover, that this is essentially the same mechanism as that underlying the photoperiodic regulation of *Arabidopsis* flowering (Suarez-Lopez et al. 2001; Imaizumi et al. 2003; Salazar et al. 2009; Song et al. 2012a, b).

Exit from winter dormancy and budflush in springtime appears to be dependent upon temperature, as chilling of dormant buds induces FT expression and development in hybrid aspen (Rinne et al. 2011).

5.6 Photoperiodic and thermal control of flowering

Reproduction in many species of plants occurs at a particular time of year in response to photoperiodic and/or thermal signals from the environment. Plants use their circadian clocks to measure photoperiod, and flowering is triggered if the day is longer than the critical day length, that is, if the plant is still receiving a light signal for a given number of hours after dawn. Photoperiodic control of flowering is mostly likely under an external coincidence mechanism governed by the presence or absence of light at a sensitive phase of the internal circadian rhythm which occurs between about 12 and 20 h after dawn (Pittendrigh and Minis 1964; Vince-Prue 1997). *Arabidopsis* is an annual plant with growth and flowering accelerated by warmth (Blazquez et al. 2003; Halliday et al. 2003). It is a facultative long-day plant, flowering in spring or summer in response to increases in day length. *Arabidopsis* flowering is delayed if plants are raised in short photoperiods, although short-day inhibition of flowering can be overcome by growth at high ambient temperatures (Balasubramanian et al. 2006).

In *Arabidopsis*, flowering is coupled to the circadian clock via the action of *GI* to promote *CO* expression in the afternoon (Mizoguchi et al. 2005). CO protein is degraded in darkness via proteasome pathways (Valverde et al. 2004; Laubinger et al. 2006). CO protein acts as a promoter of transcription of meristem identity genes such as *LEAFY* (*LFY*), *FLOWERING LOCUS T* (*FT*), and *SUPPRESSOR OF CO1* (*SOC1*), controlling the transition to flowering. Because *CO* is transcribed only in the afternoon and evening and CO is stable only in the light, the sensitive phase for daylight perception falls in the evening. Stability of CO is regulated by the phyA and cry2 photoreceptors. Another blue light-responsive protein, FLAVIN-BINDING KELCH REPEAT F-BOX PROTEIN1 (FKF1), acts indirectly as a positive regulator of *FT* transcription. *FKF1* expression is under clock control and peaks in the evening, enhancing evening CO expression specifically in long days (Imaizumi et al. 2003). In some short-day plants such as rice, Hd1 (the orthologue of CO) instead acts to inhibit transcription of Hd3 (the orthologue of FT) and hence maintains vegetative growth (Hayama and Coupland 2003).

In other species of short-day plants, such as morning glory (*Ipomoea nil*, formerly *Pharbitis nil*), the rhythm of the circadian clock controlling expression of *FT* orthologues is set by dusk, rather than dawn as in *Arabidopsis*, and expression of *IpFT1* and *IpFT2* rises only if night length exceeds 11 h (Hayama et al. 2007). Moreover, *FT* expression does not track *CO* in this species, nor does it do so in the day-neutral tomato (*Solanum lycopersicon*) (Ben-Naim et al. 2006), implying that regulation of *FT* by CO is not an invariant factor in the control of flowering.

Transport via the phloem of FT from the leaves, where the light signal is perceived, to the apical meristem is necessary for the photoperiodic induction of flowering (Corbesier et al. 2007; Jaeger and Wigge 2007). However, the thermal induction of flowering, whereby high temperature overrides a short-day signal, does not involve the photoperiodic pathway (Balasubramanian et al. 2006). The role of PIF4 in modulating growth in response to a rise in ambient temperature has been discussed already, but recent work by Wigge and colleagues has shown that PIF4 also acts on FT to promote flowering in response to high temperature. High temperature failed to induce accelerated flowering in short days in *pif4* plants, and PIF4 bound the *FT* promoter in a temperature-dependent manner, with a five-fold increase in binding at 27°C compared with 12°C, that reflected the effect of temperature upon flowering (Kumar et al. 2012).

Temperature output signals are modulated by differential incorporation of the alternative histone H2A.Z into nucleosomes, and plants unable to do this express the high-temperature transcriptome constitutively (Kumar and Wigge 2010). The proportion of H2A.Z nucleosomes at the *FT* promoter decreases at higher temperature. Moreover, plants unable to incorporate H2A.Z nucleosomes show greater binding of PIF4 to the *FT* promoter, indicating that the temperature-dependent regulation of *FT* by PIF4 is controlled at the level of chromatin accessibility (Kumar et al. 2012).

Arabidopsis accessions fall into two groups: they are either rapidly flowering summer annuals, overwintering as seeds and germinating in springtime, or vernalization-requiring winter annuals, germinating in the autumn and overwintering as rosette plants. Plants requiring vernalization must experience a long period of sustained low temperature before flowering can be induced (discussed in Chapter 4). A complex epigenetic mechanism underlies this process, centered upon the methylation state of the MADS-box protein *FLC* which binds and represses *FT* and *SOC1*, genes promoting flowering (Helliwell et al. 2006; Searle et al. 2006). During winter, cold causes enrichment of trimethylated histone H3 Lys27 in *FLC* chromatin which is required for maintenance of stable *FLC* repression (De Lucia et al. 2008). The winter-annual habit is conferred by the activation of *FLC* by *FRIGIDA* (*FRI*) (Koornneef et al. 1994; Johanson et al. 2000; Shindo et al. 2005). FRI upregulates *FLC* by interacting with the nuclear cap-binding complex (CBC), a complex which protects RNA transcripts from degradation and facilitates export to the cytoplasm (Geraldo et al. 2009). *FRI* causes an increase in *FLC* levels by enhancing histone methyltransferase (HMT) activity. At a molecular level, the summer-annual *versus* winter-annual growth habit is determined by the balance of HMT and histone demethylases acting on *FLC*; dominance of HMT produces a winter-annual plant (Ko et al. 2010). Stable repression of *FLC* following the period of vernalization requires the polycomb complex, which interacts with *FLC* via noncoding (Heo and Sung 2011). Vernalization allows plants to flower at the most appropriate time of year. They are buffered against short-term rises of temperature as low temperatures repress expression of *FLC*, but this effect is reversed on return to warm temperatures. Sustained cold is necessary to repress *FLC* stably, allowing plants to flower in spring.

5.7 Light-dependent circadian gating of cold-acclimation responses

Adaptation to cold is principally achieved by reorganization of the transcriptome which allows plants to acclimate and survive freezing temperatures that would otherwise be lethal (Kilian et al. 2007; Thomashow 2010). In response to low temperatures, plants cold acclimate by activating mechanisms which produce an increase in freezing tolerance (reviewed by Thomashow 1999, 2001, 2010). The primary event in the molecular response to cold is the induction of the CBF/DREB transcription factors, which occurs within approximately 15 min of transfer to 4°C (Stockinger et al. 1997). These proteins bind to the C-repeat/dehydration-responsive element (CRT/DRE) in genes of the CBF regulon, and target gene induction follows within 2–3 h. Plants constitutively expressing any of the CBF genes show improved freezing tolerance in the absence of a cold pretreatment. Conversely, plants with depleted CBF expression are less tolerant of freezing (Jaglo-Ottosen et al. 1998; Liu et al. 1998; Gilmour et al. 2004; Novillo et al. 2007). In natural populations of *Arabidopsis*, freezing tolerance also correlates well with CBF expression (Alonso-Blanco et al. 2005; Hannah et al. 2006). CBF-overexpressing plants display an altered growth habit, being slow-growing dwarfs with delayed flowering as overexpression of CBFs causes accumulation of DELLA proteins which restrain growth (Harberd et al. 2009) and low levels of active GAs (Achard et al. 2008).

Cold induction of CBF genes is gated by the circadian clock (Fowler and Toner 2005). Gating is a phenomenon whereby an identical stimulus provokes a different response depending on when it is received. WT *Arabidopsis* plants, for instance, are more sensitive to R light pulses during the day than they are at night (McWatters et al. 2000). Higher levels of CBF were induced following exposure to cold 4 h after lights-on (ZT 4) than 12 h later, 4 h after lights-off (ZT 16) (Fowler and Toner 2005), indicating the plant is more sensitive to temperature decreases (i.e., the gate is open) during the day than at night. This enhanced daytime sensitivity to cold may be an adaptive response as, in nature, there is a coincidence of higher temperatures and light during the day and lower temperature and dark at night.

CBF expression is itself under clock control (Harmer et al. 2000; Bieniawska et al. 2008; Dong et al. 2011) with rhythmicity being largely abolished in the *cca1lhy* mutant (Dong et al. 2011). It appears that CCA1 and LHY bind EE- and CCA1-binding sites in the CBF promoters and promote their transcription during the day. Both clock transcription factors are positive regulators of cold induction of CBFs, and, as has already been noted, cold tolerance is reduced in the *cca1lhy* mutants (Dong et al. 2011). Thus, LHY and CCA1 are necessary for freezing tolerance, with or without cold acclimation (Espinoza et al. 2010; Dong et al. 2011). Negative regulation of CBFs also appears to involve the circadian clock. Both circadian regulation of CBFs and gating of the response to cold are absent from the triple *prr5prr7prr9* mutant, but expression of CBFs in this background is consistently high (Nakamichi et al. 2009).

The EE, a nine base-pair regulatory motif identified by (Harmer et al. 2000) as important in the circadian regulation of genes due to binding by the circadian transcription factors CCA1 and LHY (Perales and Mas 2007; James et al. 2008), is overrepresented in genes induced by cold (Vogel et al. 2005; Mikkelsen and Thomashow 2009). Mutational analysis of the EE revealed it was required for expression of the cold-induced genes *CONSTANS-LIKE5* (*COL5*) and *COLD-REGULATED GENE27* (*COR27*) in partnership with ABA response element-like (ABREL) motifs (Mikkelsen and Thomashow 2009). The coupling of these two regulatory elements was enriched in cold-induced genes, and an artificial promoter containing only multiple copies of the EE and ABREL motifs was enough to induce expression in response to cold. At 24°C, an ambient temperature permissive of plant growth, cold induction of *COR27* was gated in a circadian manner and expression of both genes was regulated by the clock. Such results show the important role of these elements in reorganizing the transcriptome in response to low temperatures.

Circadian regulation of the CBF genes also involves PIF7. This binds to the G-box in the CBF2/DREB1C promoter and is required for rhythmic CBF expression but not for gating of the cold response (Kidokoro et al. 2009). PIF7 couples CBF signaling with the clock via its interactions with PHYB, the most abundant R light-sensitive phytochrome providing clock input, and TOC1, a component of the evening clock loop (Kidokoro et al. 2009). Interaction with TOC1 appears to enhance the inhibitory function of PIF7. PIF7 may therefore function as a regulator of plant growth under normal (unstressed) conditions by preventing growth retardation by the accumulation of CBF proteins. As CBF protein levels are high in the *prr5prr7prr9* mutant (Nakamichi et al. 2009), it is possible that the PRRs act upstream of PIF7. However, as PIF7 does not associate directly with any of the PRR family other than TOC1 nor with other regulators of the low-temperature response (Kidokoro et al. 2009), it is also possible that the pathways act in parallel. PIF7 was more effective at repressing CBF expression when co-expressed with phyB (Kidokoro et al. 2009), suggesting that the interaction between the active Pfr form of phyB and PIF proteins, including PIF7 (Leivar et al. 2008a), may be an important element in this response.

In a further link between light and the cold-acclimation response, low R/FR signals perceived by phytochromes can activate the CBF regulon and thus enhance freezing tolerance even in the absence of cold acclimation (Franklin and Whitelam 2007). Likewise, in tomato, overexpression of CBF1 improved resistance to cold and, under conditions of low irradiance and cold stress, increased the efficiency of photosynthesis (Zhang et al. 2011b).

5.8 Temperature and light regulation of cell membrane fatty acid composition

Other pathways beyond those discussed already are involved in plant growth responses to temperature changes. *Arabidopsis* plants without a functional *FATTY ACID DESATURASE2* (*FAD2*) gene grow normally at 22°C but do not show stem elongation at 12°C and die as a result of prolonged exposure to 6°C, well above the lethal temperature of WT plants (Miquel and Browse 1994). Other *fad* mutants also show increased sensitivity to chilling. This phenotype may result from the altered fatty acid composition of cell membranes. *fad2* mutants contain a higher proportion of saturated and partially desaturated fatty acids than WT plants (Miquel and Browse 1994; Okuley et al. 1994). *fad2* plants also show overcompensation of the circadian period, which remains constant between 12°C and 27°C, and, unlike WT plants, cannot alter membrane viscosity in temperature cycles. They do, however, show an elongated hypocotyl in response to high temperature and are early flowering regardless of temperature (Martiniere et al. 2011).

In some plant species, including cotton, there are diurnal changes in lipid composition. But in others, such as *Arabidopsis*, the proportions of the various fatty acids differ very little across a photoperiod (Martiniere et al. 2011). *Arabidopsis* plants actively manage membrane viscosity in temperature cycles by decreasing the proportion of polydesaturated fatty acids during the warm period to maintain homeostasis in viscosity (Martiniere et al. 2011). Desaturation of fatty acids makes them less able to pack tightly together and thus decreases viscosity of a membrane.

The phenotypes of *fad* mutants link thermotolerance to fatty acid composition. At 22°C, 70% of thylakoid fatty acids are trienoic, containing three double bonds, but this proportion drops at higher temperatures and this reduction in unsaturation level is associated with increased plant survival and photosynthetic function. Although at low temperatures the *fad3fad7fad8* mutant, which has no detectable trienoic fatty acids, grows poorly (possibly because of reduced recovery from photoinhibition) (Vijayan and Browse 2002; Ma and Browse 2006), at 22°C its growth and appearance are near indistinguishable from WT. It additionally showed better tolerance of high temperature (40°C) challenges and suffered less damage to PSII (Routaboul et al. 2012). Unfortunately, however, for those seeking to improve plant thermotolerance or performance by manipulation of trienoic acid levels, there is no simple trade-off whereby reducing unsaturation levels compromises growth and photosynthesis at low temperatures but improves it at high ones. In the longer term, *fad3fad7fad8* plants grown at 33°C were compromised with respect to photosynthesis, growth, and survival, indicating trienoic fatty acids are required for long-term thermotolerance (Routaboul et al. 2012).

5.9 Concluding thoughts: Implications for a changing future

Plants must continually adapt their physiology to a changing environment. Light and temperature are separate but not fully independent environmental signals encoding information about time of day and year, as well as the immediate conditions allowing (or inhibiting) photosynthesis and plant growth. The system of signal perception and transduction is complex with multiple overlapping pathways and cross talk. This is a result of the need to integrate the different types of information and coordinate the proper responses of multiple pathways. Moreover, there is a tension between the requirement to buffer the system against short-term random noise while still being sensitive to the underlying patterns of light and temperature to enable necessary physiological adjustment and adaptation. Environmental information must therefore also be integrated over time, and the circadian clock plays a key role in this process. It is clear that diverse signaling pathways converge at certain key components including PIF4, HY5, and the DELLA proteins. However, differential responses to light and temperature indicate that the downstream responses from these network control points are fine-tuned to fit the particular inputs, implying that the different signals act upon the control points in different ways. The nature of these differences is not yet fully understood.

Understanding the effect of temperature upon plant growth is not a trivial pursuit nor pursued for its own end. The Intergovernmental Panel on Climate Change (IPCC) predicts a rise in mean global temperatures of up to 6.4°C by 2100 (IPCC 2007). This poses a potentially major problem for food security and a challenge for global agriculture (discussed in detail in Chapter 8). In the cases of major crop cereals, wheat, maize, and barley, there is a negative correlation between yield and increased temperature, and recent warming has contributed to an estimated global loss of $5 billion for these crops alone (Lobell and Field 2007). Even isolated episodes of high temperature may have a severe impact upon final yield if they occur at a sensitive period in plant development. For instance, high temperature during anther development significantly reduces cereal yield by inducing male sterility (Abiko et al. 2005; Barnabas et al. 2008; Sakata et al. 2010). Responses to climate change may have to include extending the margin of cultivation northwards while abandoning southern areas currently under production, as is predicted will occur with wheat farming in Canada and the USA (Ortiz et al. 2008). It will also mean the development of new varieties able to thrive under the warmer and drier conditions of the future. Past experience may provide an insight into the future: between 1839 and 1929, development of new varieties of hard winter wheat allowed the center of North American wheat production to be pushed into areas once considered too arid and too harsh for cultivation (Olmstead and Rhode 2011). Modeling approaches that take into account plant responses to high temperature may assist conventional breeding programs in developing new cultivars suitable for future environments. As an example of this approach, a predictive model of wheat yield has suggested that the greatest challenge for wheat farming in England and Wales will not be drought stress, since increased winter rainfall will mitigate the effects of drier summers, but heat stress due to male sterility (Semenov and Shewry 2011). If this is so then, since the sterility is due in large part to a drop in endogenous auxin levels, targeting auxin biosynthesis pathways (by, e.g., increasing expression of *YUC8* at high temperature) or application of exogenous auxin may ameliorate the effects of heat (Sakata et al. 2010).

Conservation of pathways between plant species will allow transfer of knowledge gained from the model plant *Arabidopsis* to economically important crop species. The molecular networks uncovered in *Arabidopsis* will therefore underpin agricultural developments. Nowhere is this more so than in modeling, as genetically informed models allow predictions of life history and yield in crop plants in real-world conditions (Wilczek et al. 2010). Recent work has used field site data on flowering times of *Arabidopsis* accessions and mutants grown in a transect across Europe to develop a phenological model of flowering that utilized knowledge of genetic pathways and the local patterns of photoperiodic and temperature change (Wilczek et al. 2009). Later iterations of this model added in the effects of night temperature and phytochrome mutants, feeding back into empirical work to implicate phyB in the temporal gating of temperature-induced flowering (Chew et al. 2012).

In face of climate change, understanding the processes of plant adaptation to a dynamic environment is an essential exercise that will provide a strong foundation for the development of the novel crop varieties needed to sustain human populations in the time ahead. Future work is needed to determine the nature of the temperature receptors of higher plants and the manner in which they are connected to photoreceptors. Another fruitful area is collaboration between theoreticians and experimentalists in developing and testing models based upon transcriptomic and proteomic data and data collected from the field. The predictive power of modeling can allow better understanding of molecular signaling pathways (Salazar et al. 2009). Such work can inform future research in a virtuous cycle from computer to laboratory to field.

References

Abiko M, Akibayashi K, Sakata T, Kimura M, Kihara M, Itoh K, Asamizu E, Sato S, Takahashi H, Higashitani A (2005) High temperature induction of male sterility during barley (*Hordeum vulgare* L.) anther development is mediated by transcriptional inhibition. Sex. Plant Reprod. 18: 91–100.

Achard P, Liao L, Jiang C, Desnos T, Bartlett J, Fu X, Harberd NP (2007) DELLAs contribute to plant photomorphogenesis. Plant Physiol. 143: 1163–1172.

Achard P, Gong F, Cheminant S, Alioua M, Hedden P, Genschik P (2008) The cold-inducible CBF1 factor-dependent signaling pathway modulates the accumulation of the growth-repressing DELLA proteins via its effect on gibberellin metabolism. Plant Cell 20: 2117–2129.

Al-Sady B, Ni W, Kircher S, Schafer E, Quail PH (2006) Photoactivated phytochrome induces rapid PIF3 phosphorylation prior to proteasome-mediated degradation. Mol. Cell 23: 439–446.

Al-Sady B, Kikis EA, Monte E, Quail PH (2008) Mechanistic duality of transcription factor function in phytochrome signaling. Proc. Natl. Acad. Sci. USA 105: 2232–2237.

Alabadi D, Blazquez MA (2008) Integration of light and hormone signals. Plant Signal. Behav. 3: 448–449.

Alonso-Blanco C, Gomez-Mena C, Llorente F, Koornneef M, Salinas J, Martinez-Zapater JM (2005) Genetic and molecular analyses of natural variation indicate CBF2 as a candidate gene for underlying a freezing tolerance quantitative trait locus in *Arabidopsis*. Plant Physiol. 139: 1304–1312.

Andronis C, Barak S, Knowles SM, Sugano S, Tobin EM (2008) The clock protein CCA1 and the bZIP transcription factor HY5 physically interact to regulate gene expression in *Arabidopsis*. Mol. Plant 1: 58–67.

Ang L-H, Deng X-W (1994) Regulatory hierarchy of photo-morphogenic loci: Allele-specific and light-dependent interaction between the HY5 and COP1 loci. Plant Cell 6: 613–628.

Aschoff J (1979) Circadian rhythms: Influences of internal and external factors on the period measured in constant conditions. Z. Tierpsychol. 49: 225–249.

Atkin OK, Loveys BR, Atkinson LJ, Pons TL (2006) Phenotypic plasticity and growth temperature: Understanding interspecific variability. J. Exp. Bot. 57: 267–281.

Balasubramanian S, Sureshkumar S, Lempe J, Weigel D (2006) Potent induction of *Arabidopsis thaliana* flowering by elevated growth temperature. PLoS Genet. 2: e106.

Barnabas B, Jager K, Feher A (2008) The effect of drought and heat stress on reproductive processes in cereals. Plant Cell Environ. 31: 11–38.

Battisti DS, Naylor RL (2009) Historical warnings of future food insecurity with unprecedented seasonal heat. Science 323: 240–244.

Ben-Naim O, Eshed R, Parnis A, Teper-Bamnolker P, Shalit A, Coupland G, Samach A, Lifschitz E (2006) The CCAAT binding factor can mediate interactions between CONSTANS-like proteins and DNA. Plant J. 46: 462–476.

Bieniawska Z, Espinoza C, Schlereth A, Sulpice R, Hincha DK, Hannah MA (2008) Disruption of the *Arabidopsis* circadian clock is responsible for extensive variation in the cold-responsive transcriptome. Plant Physiol. 147: 263–279.

Blasing OE, Gibon Y, Gunther M, Hohne M, Morcuende R, Osuna D, Thimm O, Usadel B, Scheible WR, Stitt M (2005) Sugars and circadian regulation make major contributions to the global regulation of diurnal gene expression in *Arabidopsis*. Plant Cell 17: 3257–3281.

Blazquez MA, Ahn JH, Weigel D (2003) A thermosensory pathway controlling flowering time in *Arabidopsis thaliana*. Nat. Genet. 33: 168–171.

Burning E (1936) Die engogene Tagersrhythmik als Grunderlage der photoperiodischen Reaktion. Ber. Dtsch. Bot. Ges. 54: 590–607.

Capel J, Jarillo JA, Madueno F, Jorquera MJ, Martinez-Zapater JM, Salinas J (1998) Low temperature regulates *Arabidopsis* Lhcb gene expression in a light-independent manner. Plant J. 13: 411–418.

Casal JJ (2012) Shade avoidance. *Arabidopsis* Book. 10: e0157. American Society of Plant Biologist.

Castillon A, Shen H, Huq E (2007) Phytochrome interacting factors: Central players in phytochrome-mediated light signaling networks. Trends Plant Sci. 12: 514–521.

Catala R, Medina J, Salinas J (2011) Integration of low temperature and light signaling during cold acclimation response in *Arabidopsis*. Proc. Natl. Acad. Sci. USA 108: 16475–16480.

Chen H, Xiong L (2008) Role of HY5 in abscisic acid response in seeds and seedlings. Plant Signal. Behav. 3: 986–988.

Chew YH, Wilczek AM, Williams M, Welch SM, Schmitt J, Halliday KJ (2012) An augmented *Arabidopsis* phenology model reveals seasonal temperature control of flowering time. New Phytol. 194: 654–665.

Corbesier L, Vincent C, Jang S, Fornara F, Fan Q, Searle I, Giakountis A, Farrona S, Gissot L, Turnbull C, Coupland G (2007) FT protein movement contributes to long-distance signaling in floral induction of *Arabidopsis*. Science 316: 1030–1033.

Covington MF, Panda S, Liu XL, Strayer CA, Wagner DR, Kay SA (2001) ELF3 modulates resetting of the circadian clock in *Arabidopsis*. Plant Cell 13: 1305–1315.

Daniel X, Sugano S, Tobin EM (2004) CK2 phosphorylation of CCA1 is necessary for its circadian oscillator function in *Arabidopsis*. Proc. Natl. Acad. Sci. USA 101: 3292–3297.

De Lucia F, Crevillen P, Jones AM, Greb T, Dean C (2008) A PHD-polycomb repressive complex 2 triggers the epigenetic silencing of FLC during vernalization. Proc. Natl. Acad. Sci. USA 105: 16831–16836.

Devlin PF, Kay SA (2000) Cryptochromes are required for phytochrome signaling to the circadian clock but not for rhythmicity. Plant Cell 12: 2499–2510.

Dixon LE, Knox K, Kozma-Bognar L, Southern MM, Pokhilko A, Millar AJ (2011) Temporal repression of core circadian genes is mediated through EARLY FLOWERING 3 in *Arabidopsis*. Curr. Biol. 21: 120–125.

Dodd AN, Salathia N, Hall A, Kevei E, Toth R, Nagy F, Hibberd JM, Millar AJ, Webb AA (2005) Plant circadian clocks increase photosynthesis, growth, survival, and competitive advantage. Science 309: 630–633.

Dong MA, Farre EM, Thomashow MF (2011) Circadian clock-associated 1 and late elongated hypocotyl regulate expression of the C-repeat binding factor (CBF) pathway in *Arabidopsis*. Proc. Natl. Acad. Sci. USA 108: 7241–7246.

Duek PD, Fankhauser C (2003) HFR1, a putative bHLH transcription factor, mediates both phytochrome A and cryptochrome signalling. Plant J. 34: 827–836.

Edwards KD, Lynn JR, Gyula P, Nagy F, Millar AJ (2005) Natural allelic variation in the temperature-compensation mechanisms of the *Arabidopsis thaliana* circadian clock. Genetics 170: 387–400.

Edwards KD, Anderson PE, Hall A, Salathia NS, Locke JC, Lynn JR, Straume M, Smith JQ, Millar AJ (2006) FLOWERING LOCUS C mediates natural variation in the high-temperature response of the *Arabidopsis* circadian clock. Plant Cell 18: 639–650.

Espinoza C, Bieniawska Z, Hincha D, Hannah M (2008) Interactions between the circadian clock and cold-response in *Arabidopsis*. Plant Signal. Behav. 3: 593–594.

Espinoza C, Degenkolbe T, Caldana C, Zuther E, Leisse A, Willmitzer L, Hincha DK, Hannah MA (2010) Interaction with diurnal and circadian regulation results in dynamic metabolic and transcriptional changes during cold acclimation in *Arabidopsis*. PLoS One 5: e14101.

Fairchild CD, Schumaker MA, Quail PH (2000) HFR1 encodes an atypical bHLH protein that acts in phytochrome A signal transduction. Genes Dev. 14: 2377–2391.

Fankhauser C, Chory J (2000) RSF1, an *Arabidopsis* locus implicated in phytochrome A signaling. Plant Physiol. 124: 39–45.

Favory JJ, Stec A, Gruber H, Rizzini L, Oravecz A, Funk M, Albert A, Cloix C, Jenkins GI, Oakeley EJ, Seidlitz HK, Nagy F, Ulm R (2009) Interaction of COP1 and UVR8 regulates UV-B-induced photomorphogenesis and stress acclimation in *Arabidopsis*. EMBO J. 28: 591–601.

Feng S, Martinez C, Gusmaroli G, Wang Y, Zhou J, Wang F, Chen L, Yu L, Iglesias-Pedraz JM, Kircher S, Schafer E, Fu X, Fan LM, Deng XW (2008) Coordinated regulation of *Arabidopsis thaliana* development by light and gibberellins. Nature 451: 475–479.

Fitter AH, Fitter RS (2002) Rapid changes in flowering time in British plants. Science 296: 1689–1691.

Foreman J, Johansson H, Hornitschek P, Josse EM, Fankhauser C, Halliday KJ (2011) Light receptor action is critical for maintaining plant biomass at warm ambient temperatures. Plant J. 65: 441–452.

Fowler A, Toner M (2005) Cryo-injury and biopreservation. Ann. N. Y. Acad. Sci. 1066: 119–135.

Franklin KA (2008) Shade avoidance. New Phytol. 179: 930–944.

Franklin KA, Whitelam GC (2007) Light-quality regulation of freezing tolerance in *Arabidopsis thaliana*. Nat. Genet. 39: 1410–1413.

Franklin KA, Praekelt U, Stoddart WM, Billingham OE, Halliday KJ, Whitelam GC (2003) Phytochromes B, D, and E act redundantly to control multiple physiological responses in *Arabidopsis*. Plant Physiol. 131: 1340–1346.

Franklin KA, Lee SH, Patel D, Kumar SV, Spartz AK, Gu C, Ye S, Yu P, Breen G, Cohen JD, Wigge PA, Gray WM (2011) Phytochrome-interacting factor 4 (PIF4) regulates auxin biosynthesis at high temperature. Proc. Natl. Acad. Sci. USA 108: 20231–20235.

Gallego-Bartolome J, Alabadi D, Blazquez MA (2011) DELLA-induced early transcriptional changes during etiolated development in *Arabidopsis thaliana*. PLoS One 6: e23918.

Geraldo N, Baurle I, Kidou S, Hu X, Dean C (2009) FRIGIDA delays flowering in *Arabidopsis* via a cotranscriptional mechanism involving direct interaction with the nuclear cap-binding complex. Plant Physiol. 150: 1611–1618.

Gibon Y, Blasing OE, Palacios-Rojas N, Pankovic D, Hendriks JH, Fisahn J, Hohne M, Gunther M, Stitt M (2004) Adjustment of diurnal starch turnover to short days: Depletion of sugar during the night leads to a temporary inhibition of carbohydrate utilization, accumulation of sugars and post-translational activation of ADP-glucose pyrophosphorylase in the following light period. Plant J. 39: 847–862.

Gilmour SJ, Fowler SG, Thomashow MF (2004) *Arabidopsis* transcriptional activators CBF1, CBF2, and CBF3 have matching functional activities. Plant Mol. Biol. 54: 767–781.

Gong W, He K, Covington M, Dinesh-Kumar SP, Snyder M, Harmer SL, Zhu YX, Deng XW (2008) The development of protein microarrays and their applications in DNA–protein and protein–protein interaction analyses of *Arabidopsis* transcription factors. Mol. Plant 1: 27–41.

Gould PD, Locke JC, Larue C, Southern MM, Davis SJ, Hanano S, Moyle R, Milich R, Putterill J, Millar AJ, Hall A (2006) The molecular basis of temperature compensation in the *Arabidopsis* circadian clock. Plant Cell 18: 1177–1187.

Gray WM, Ostin A, Sandberg G, Romano CP, Estelle M (1998) High temperature promotes auxin-mediated hypocotyl elongation in *Arabidopsis*. Proc. Natl. Acad. Sci. USA 95: 7197–7202.

Gruber H, Heijde M, Heller W, Albert A, Seidlitz HK, Ulm R (2010) Negative feedback regulation of UV-B-induced photomorphogenesis and stress acclimation in *Arabidopsis*. Proc. Natl. Acad. Sci. USA 107: 20132–20137.

Halliday KJ, Whitelam GC (2003) Changes in photoperiod or temperature alter the functional relationships between phytochromes and reveal roles for phyD and phyE. Plant Physiol. 131: 1913–1920.

Halliday KJ, Salter MG, Thingnaes E, Whitelam GC (2003) Phytochrome control of flowering is temperature sensitive and correlates with expression of the floral integrator FT. Plant J. 33: 875–885.

Hannah MA, Wiese D, Freund S, Fiehn O, Heyer AG, Hincha DK (2006) Natural genetic variation of freezing tolerance in *Arabidopsis*. Plant Physiol. 142: 98–112.

Harberd NP (2003) Botany. Relieving DELLA restraint. Science 299: 1853–1854.

Harberd NP, King KE, Carol P, Cowling RJ, Peng J, Richards DE (1998) Gibberellin: Inhibitor of an inhibitor of…? Bioessays 20: 1001–1008.

Harberd NP, Belfield E, Yasumura Y (2009) The angiosperm gibberellin-GID1-DELLA growth regulatory mechanism: How an "inhibitor of an inhibitor" enables flexible response to fluctuating environments. Plant Cell 21: 1328–1339.

Harmer S (2009) The circadian system. Ann. Rev. Plant Biol. 60: 357–377.

Harmer SL, Hogenesch JB, Straume M, Chang HS, Han B, Zhu T, Wang X, Kreps JA, Kay SA (2000) Orchestrated transcription of key pathways in *Arabidopsis* by the circadian clock. Science 290: 2110–2113.

Hayama R, Coupland G (2003) Shedding light on the circadian clock and the photoperiodic control of flowering. Curr. Opin. Plant Biol. 6: 13–19.

Hayama R, Agashe B, Luley E, King R, Coupland G (2007) A circadian rhythm set by dusk determines the expression of FT homologs and the short-day photoperiodic flowering response in Pharbitis. Plant Cell 19: 2988–3000.

Helfer A, Nusinow DA, Chow BY, Gehrke AR, Bulyk ML, Kay SA (2011) LUX ARRHYTHMO encodes a nighttime repressor of circadian gene expression in the *Arabidopsis* core clock. Curr. Biol. 21: 126–133.

Helliwell CA, Wood CC, Robertson M, Peacock WJ, Dennis ES (2006) The *Arabidopsis* FLC protein interacts directly in vivo with SOC1 and FT chromatin and is part of a high-molecular-weight protein complex. Plant J. 46: 183–192.

Heo JB, Sung S (2011) Vernalization-mediated epigenetic silencing by a long intronic noncoding RNA. Science 331: 76–79.

Heschel MS, Selby J, Butler C, Whitelam GC, Sharrock RA, Donohue K (2007) A new role for phytochromes in temperature-dependent germination. New Phytol. 174: 735–741.

Hornitschek P, Lorrain S, Zoete V, Michielin O, Fankhauser C (2009) Inhibition of the shade avoidance response by formation of non-DNA binding bHLH heterodimers. EMBO J. 28: 3893–3902.

Hornitschek P, Kohnen MV, Lorrain S, Rougemont J, Ljung K, Lopez-Vidriero I, Franco-Zorrilla JM, Solano R, Trevisan M, Pradervand S, Xenarios I, Fankhauser C (2012) Phytochrome interacting factors 4 and 5 control seedling growth in changing light conditions by directly controlling auxin signaling. Plant J. 71: 699–711.

Huq E, Al-Sady B, Hudson M, Kim C, Apel K, Quail PH (2004) Phytochrome-interacting factor 1 is a critical bHLH regulator of chlorophyll biosynthesis. Science 305: 1937–1941.

Ibanez C, Ramos A, Acebo P, Contreras A, Casado R, Allona I, Aragoncillo C (2008) Overall alteration of circadian clock gene expression in the chestnut cold response. PLoS One 3: e3567.

Ibanez C, Kozarewa I, Johansson M, Ogren E, Rohde A, Eriksson ME (2010) Circadian clock components regulate entry and affect exit of seasonal dormancy as well as winter hardiness in Populus trees. Plant Physiol. 153: 1823–1833.

Intergovernmental Panel on Climate Change (IPCC) (2007) The physical science basis. In: Solomon S, Qin D, Manning M, Chen Z, Marquis M, Averyt M, Tignor M, Miller M (eds). Contribution of Working Group I to the Fourth Assessment Report of the Intergovernmental Panel on Climate Change. Cambridge University Press, Cambridge.

Imaizumi T, Tran HG, Swartz TE, Briggs WR, Kay SA (2003) FKF1 is essential for photoperiodic-specific light signalling in *Arabidopsis*. Nature 426: 302–306.

Jaeger KE, Wigge PA (2007) FT protein acts as a long-range signal in *Arabidopsis*. Curr. Biol. 17: 1050–1054.

Jaglo-Ottosen KR, Gilmour SJ, Zarka DG, Schabenberger O, Thomashow MF (1998) *Arabidopsis* CBF1 overexpression induces COR genes and enhances freezing tolerance. Science 280: 104–106.

James AB, Monreal JA, Nimmo GA, Kelly CL, Herzyk P, Jenkins GI, Nimmo HG (2008) The circadian clock in *Arabidopsis* roots is a simplified slave version of the clock in shoots. Science 322: 1832–1835.

James AB, Syed NH, Bordage S, Marshall J, Nimmo GA, Jenkins GI, Herzyk P, Brown JW, Nimmo HG (2012) Alternative splicing mediates responses of the *Arabidopsis* circadian clock to temperature changes. Plant Cell 24: 961–981.

Johanson U, West J, Lister C, Michaels S, Amasino R, Dean C (2000) Molecular analysis of FRIGIDA, a major determinant of natural variation in *Arabidopsis* flowering time. Science 290: 344–347.

Karlberg A, Bako L, Bhalerao RP (2011) Short day-mediated cessation of growth requires the downregulation of AINTEGUMENTALIKE1 transcription factor in hybrid aspen. PLoS Genet. 7: e1002361.

Khanna R, Huq F, Kikis EA, Al-Sady B, Lanzatella C, Quail PH (2004) A novel molecular recognition motif necessary for targeting photoactivated phytochrome signaling to specific basic helix–loop–helix transcription factors. Plant Cell 16: 3033–3044.

Kidokoro S, Maruyama K, Nakashima K, Imura Y, Narusaka Y, Shinwari ZK, Osakabe Y, Fujita Y, Mizoi J, Shinozaki K, Yamaguchi-Shinozaki K (2009) The phytochrome-interacting factor PIF7 negatively regulates DREB1 expression under circadian control in *Arabidopsis*. Plant Physiol. 151: 2046–2057.

Kilian J, Whitehead D, Horak J, Wanke D, Weinl S, Batistic O, D'Angelo C, Bornberg-Bauer E, Kudla J, Harter K (2007) The AtGenExpress global stress expression data set: Protocols, evaluation and model data analysis of UV-B light, drought and cold stress responses. Plant J. 50: 347–363.

Kim J, Kim Y, Yeom M, Kim JH, Nam HG (2008) FIONA1 is essential for regulating period length in the *Arabidopsis* circadian clock. Plant Cell 20: 307–319.

Ko JH, Mitina I, Tamada Y, Hyun Y, Choi Y, Amasino RM, Noh B, Noh YS (2010) Growth habit determination by the balance of histone methylation activities in *Arabidopsis*. EMBO J. 29: 3208–3215.

Koini MA, Alvey L, Allen T, Tilley CA, Harberd NP, Whitelam GC, Franklin KA (2009) High temperature-mediated adaptations in plant architecture require the bHLH transcription factor PIF4. Curr. Biol. 19: 408–413.

Kolmos E, Nowak M, Werner M, Fischer K, Schwarz G, Mathews S, Schoof H, Nagy F, Bujnicki JM, Davis SJ (2009) Integrating ELF4 into the circadian system through combined structural and functional studies. HFSP J. 3: 350–366.

Koornneef M, Blankestidjn-deVries H, Hanhart C, Soppe W, Peeters T (1994) The phenotype of some late flowering mutants is enhanced by a locus on chromosome 5 that is not effective in the Landsberg erecta wild-type. Plant J. 6: 911–919.

Kozarewa I, Ibanez C, Johansson M, Ogren E, Mozley D, Nylander E, Chono M, Moritz T, Eriksson ME (2010) Alteration of PHYA expression change circadian rhythms and timing of bud set in Populus. Plant Mol. Biol. 73: 143–156.

Kumar SV, Wigge PA (2010) H2A.Z-containing nucleosomes mediate the thermosensory response in *Arabidopsis*. Cell 140: 136–147.

Kumar SV, Lucyshyn D, Jaeger KE, Alos E, Alvey E, Harberd NP, Wigge PA (2012) Transcription factor PIF4 controls the thermosensory activation of flowering. Nature 484: 242–245.

Lau OS, Deng XW (2010) Plant hormone signaling lightens up: Integrators of light and hormones. Curr. Opin. Plant Biol. 13: 571–577.

Laubinger S, Marchal V, Le Gourrierec J, Wenkel S, Adrian J, Jang S, Kulajta C, Braun H, Coupland G, Hoecker U (2006) *Arabidopsis* SPA proteins regulate photoperiodic flowering and interact with the floral inducer CONSTANS to regulate its stability. Development 133: 3213–3222.

Lee J, He K, Stolc V, Lee H, Figueroa P, Gao Y, Tongprasit W, Zhao H, Lee I, Deng XW (2007) Analysis of transcription factor HY5 genomic binding sites revealed its hierarchical role in light regulation of development. Plant Cell 19: 731–749.

Leivar P, Quail PH (2011) PIFs: Pivotal components in a cellular signaling hub. Trends Plant Sci. 16: 19–28.

Leivar P, Monte E, Al-Sady B, Carle C, Storer A, Alonso JM, Ecker JR, Quail PH (2008a) The *Arabidopsis* phytochrome-interacting factor PIF7, together with PIF3 and PIF4, regulates responses to prolonged red light by modulating phyB levels. Plant Cell 20: 337–352.

Leivar P, Monte E, Oka Y, Liu T, Carle C, Castillon A, Huq E, Quail PH (2008b) Multiple phytochrome-interacting bHLH transcription factors repress premature seedling photomorphogenesis in darkness. Curr. Biol. 18: 1815–1823.

Leivar P, Monte E, Cohn MM, Quail PH (2012) Phytochrome signaling in green *Arabidopsis* seedlings: Impact assessment of a mutually negative phyB-PIF feedback loop. Mol. Plant 5: 734–749.

Li J, Li G, Gao S, Martinez C, He G, Zhou Z, Huang X, Lee JH, Zhang H, Shen Y, Wang H, Deng XW (2010) *Arabidopsis* transcription factor ELONGATED HYPOCOTYL5 plays a role in the feedback regulation of phytochrome A signaling. Plant Cell 22: 3634–3649.

Li Z, Zhang L, Yu Y, Quan R, Zhang Z, Zhang H, Huang R (2011) The ethylene response factor AtERF11 that is transcriptionally modulated by the bZIP transcription factor HY5 is a crucial repressor for ethylene biosynthesis in *Arabidopsis*. Plant J. 68: 88–99.

Li L, Ljung K, Breton G, Schmitz RJ, Pruneda-Paz J, Cowing-Zitron C, Cole BJ, Ivans LJ, Pedmale UV, Jung HS, Ecker JR, Kay SA, Chory J (2012) Linking photoreceptor excitation to changes in plant architecture. Genes Dev. 26: 785–790.

Liu Q, Kasuga M, Sakuma Y, Abe H, Miura S, Yamaguchi-Shinozaki K, Shinozaki K (1998) Two transcription factors, DREB1 and DREB2, with an EREBP/AP2 DNA binding domain separate two cellular signal transduction pathways in drought- and low-temperature-responsive gene expression, respectively, in *Arabidopsis*. Plant Cell 10: 1391–1406.

Liu B, Zuo Z, Liu H, Liu X, Lin C (2011) *Arabidopsis* cryptochrome 1 interacts with SPA1 to suppress COP1 activity in response to blue light. Genes Dev. 25: 1029–1034.

Lobell D, Field C (2007) Global scale climate-crop yield relationships and the impact of recent warming. Environ. Res. Lett. 2: 1–7.

Lorrain S, Allen T, Duek PD, Whitelam GC, Fankhauser C (2008) Phytochrome-mediated inhibition of shade avoidance involves degradation of growth-promoting bHLH transcription factors. Plant J. 53: 312–323.

Lorrain S, Trevisan M, Pradervand S, Fankhauser C (2009) Phytochrome interacting factors 4 and 5 redundantly limit seedling de-etiolation in continuous far-red light. Plant J. 60: 449–461.

de Lucas M, Daviere JM, Rodriguez-Falcon M, Pontin M, Iglesias-Pedraz JM, Lorrain S, Fankhauser C, Blazquez MA, Titarenko E, Prat S (2008) A molecular framework for light and gibberellin control of cell elongation. Nature 451: 480–484.

Ma X, Browse J (2006) Altered rates of protein transport in *Arabidopsis* mutants deficient in chloroplast membrane unsaturation. Phytochemistry 67: 1629–1636.

Martiniere A, Shvedunova M, Thomson AJ, Evans NH, Penfield S, Runions J, McWatters HG (2011) Homeostasis of plasma membrane viscosity in fluctuating temperatures. New Phytol. 192: 328–337.

Mas P, Yanovsky MJ (2009) Time for circadian rhythms: Plants get synchronized. Curr. Opin. Plant Biol. 12: 574–579.

Matsushika A, Murakami M, Ito S, Nakamichi N, Yamashino T, Mizuno T (2007) Characterization of circadian-associated pseudo-response regulators: I. Comparative studies on a series of transgenic lines misexpressing five distinctive PRR genes in *Arabidopsis thaliana*. Biosci. Biotechnol. Biochem. 71: 527–534.

McClung CR, Davis SJ (2010) Ambient thermometers in plants: From physiological outputs towards mechanisms of thermal sensing. Curr. Biol. 20: R1086–R1092.

McWatters HG, Devlin PF (2011) Timing in plants – A rhythmic arrangement. FEBS Lett. 585: 1474–1484.

McWatters HG, Bastow RM, Hall A, Millar AJ (2000) The ELF3 zeitnehmer regulates light signalling to the circadian clock. Nature 408: 716–720.

Mehra A, Shi M, Baker CL, Colot HV, Loros JJ, Dunlap JC (2009) A role for casein kinase 2 in the mechanism underlying circadian temperature compensation. Cell 137: 749–760.

Michael TP, Breton G, Hazen SP, Priest H, Mockler TC, Kay SA, Chory J (2008) A morning-specific phytohormone gene expression program underlying rhythmic plant growth. PLoS Biol. 6: e225.

Mikkelsen MD, Thomashow MF (2009) A role for circadian evening elements in cold-regulated gene expression in *Arabidopsis*. Plant J. 60: 328–339.

Miquel MF, Browse JA (1994) High-oleate oilseeds fail to develop at low temperature. Plant Physiol. 106: 421–427.

Mizoguchi T, Wright L, Fujiwara S, Cremer F, Lee K, Onouchi H, Mouradov A, Fowler S, Kamada H, Putterill J, Coupland G (2005) Distinct roles of GIGANTEA in promoting flowering and regulating circadian rhythms in *Arabidopsis*. Plant Cell 17: 2255–2270.

Mizuno T, Nakamichi N (2005) Pseudo-Response Regulators (PRRs) or True Oscillator Components (TOCs). Plant Cell Physiol. 46: 677–685.

Nagy F, Schafer E (2002) Phytochromes control photomorphogenesis by differentially regulated, interacting signaling pathways in higher plants. Annu. Rev. Plant Biol. 53: 329–355.

Nakamichi N, Kita M, Ito S, Sato E, Yamashino T, Mizuno T (2005) The *Arabidopsis* pseudo-response regulators, PRR5 and PRR7, coordinately play essential roles for circadian clock function. Plant Cell Physiol. 46: 609–619.

Nakamichi N, Kita M, Niinuma K, Ito S, Yamashino T, Mizoguchi T, Mizuno T (2007) *Arabidopsis* clock-associated pseudo-response regulators PRR9, PRR7 and PRR5 coordinately and positively regulate flowering time through the canonical CONSTANS-dependent photoperiodic pathway. Plant Cell Physiol. 48: 822–832.

Nakamichi N, Kusano M, Fukushima A, Kita M, Ito S, Yamashino T, Saito K, Sakakibara H, Mizuno T (2009) Transcript profiling of an *Arabidopsis* PSEUDO RESPONSE REGULATOR arrhythmic triple mutant reveals a role for the circadian clock in cold stress response. Plant Cell Physiol. 50: 447–462.

Ni M, Tepperman JM, Quail PH (1999) Binding of phytochrome B to its nuclear signalling partner PIF3 is reversibly induced by light. Nature 400: 781–784.

Novillo F, Medina J, Salinas J (2007) *Arabidopsis* CBF1 and CBF3 have a different function than CBF2 in cold acclimation and define different gene classes in the CBF regulon. Proc. Natl. Acad. Sci. USA 104: 21002–21007.

Nozue K, Covington MF, Duek PD, Lorrain S, Fankhauser C, Harmer SL, Maloof JN (2007) Rhythmic growth explained by coincidence between internal and external cues. Nature 448: 358–361.

Nozue K, Harmer SL, Maloof JN (2011) Genomic analysis of circadian clock-, light-, and growth-correlated genes reveals PHYTOCHROME-INTERACTING FACTOR5 as a modulator of auxin signaling in *Arabidopsis*. Plant Physiol. 156: 357–372.

Nusinow DA, Helfer A, Hamilton EE, King JJ, Imaizumi T, Schultz TF, Farré EM, Kay SA (2011) The ELF4–ELF3–LUX complex links the circadian clock to diurnal control of hypocotyl growth. Nature 475: 398–402.

Okuley J, Lightner J, Feldmann K, Yadav N, Lark E, Browse J (1994) *Arabidopsis* FAD2 gene encodes the enzyme that is essential for polyunsaturated lipid synthesis. Plant Cell 6: 147–158.

Olmstead AL, Rhode PW (2011) Adapting North American wheat production to climatic challenges, 1839–2009. Proc. Natl. Acad. Sci. USA 108: 480–485.

Olsen JE, Junttila O (2002) Far red end-of-day treatment restores wild type-like plant length in hybrid aspen overexpressing phytochrome A. Physiol. Plant. 115: 448–457.

Onai K, Ishiura M (2005) PHYTOCLOCK 1 encoding a novel GARP protein essential for the *Arabidopsis* circadian clock. Genes Cells 10: 963–972.

Ortiz R, Sayre K, Govaerts B, Gupta R, Subbarao G, Ban T, Hodson D, Dixon J, Ivan Ortiz Monsterio J, Reynolds M (2008) Climate change: Can wheat beat the heat? Agric. Ecosyst. Environ. 126: 46–58.

Osterlund MT, Hardtke CS, Wei N, Deng XW (2000) Targeted destabilization of HY5 during light-regulated development of *Arabidopsis*. Nature 405: 462–466.

Park E, Kim J, Lee Y, Shin J, Oh E, Chung WI, Liu JR, Choi G (2004) Degradation of phytochrome interacting factor 3 in phytochrome-mediated light signaling. Plant Cell Physiol. 45: 968–975.

Patel D, Franklin KA (2009) Temperature-regulation of plant architecture. Plant Signal. Behav. 4: 577–579.

Penfield S, Josse EM, Kannangara R, Gilday AD, Halliday KJ, Graham IA (2005) Cold and light control seed germination through the bHLH transcription factor SPATULA. Curr. Biol. 15: 1998–2006.

Perales M, Mas P (2007) A functional link between rhythmic changes in chromatin structure and the *Arabidopsis* biological clock. Plant Cell 19: 2111–2123.

Pittendrigh CS (1954) On the temperature independence in the clock system controlling emergence time in *Drosophila*. Proc. Natl. Acad. Sci. USA 40: 1018–1029.

Pittendrigh C, Minis D (1964) The entrainment of the circadian oscillations by light and their role as photoperiodic clocks. Am. Nat. 98: 261–294.

Pokhilko A, Fernandez AP, Edwards KD, Southern MM, Halliday KJ, Millar AJ (2012) The clock gene circuit in *Arabidopsis* includes a repressilator with additional feedback loops. Mol. Syst. Biol. 8: 574.

Portoles S, Mas P (2010) The functional interplay between protein kinase CK2 and CCA1 transcriptional activity is essential for clock temperature compensation in *Arabidopsis*. PLoS Genet. 6: e1001201.

Ramos A, Perez-Solis E, Ibanez C, Casado R, Collada C, Gomez L, Aragoncillo C, Allona I (2005) Winter disruption of the circadian clock in chestnut. Proc. Natl. Acad. Sci. USA 102: 7037–7042.

Rawat R, Takahashi N, Hsu PY, Jones MA, Schwartz J, Salemi MR, Phinney BS, Harmer SL (2011) REVEILLE8 and PSEUDO-REPONSE REGULATOR5 form a negative feedback loop within the *Arabidopsis* circadian clock. PLoS Genet. 7: e1001350.

Rinne PL, Welling A, Vahala J, Ripel L, Ruonala R, Kangasjarvi J, van der Schoot C (2011) Chilling of dormant buds hyperinduces FLOWERING LOCUS T and recruits GA-inducible 1,3-beta-glucanases to reopen signal conduits and release dormancy in Populus. Plant Cell 23: 130–146.

Routaboul JM, Skidmore C, Wallis JG, Browse J (2012) *Arabidopsis* mutants reveal that short- and long-term thermotolerance have different requirements for trienoic fatty acids. J. Exp. Bot. 63: 1435–1443.

Saijo Y, Sullivan JA, Wang H, Yang J, Shen Y, Rubio V, Ma L, Hoecker U, Deng XW (2003) The COP1–SPA1 interaction defines a critical step in phytochrome A-mediated regulation of HY5 activity. Genes Dev. 17: 2642–2647.

Sakata T, Oshino T, Miura S, Tomabechi M, Tsunaga Y, Higashitani N, Miyazawa Y, Takahashi H, Watanabe M, Higashitani A (2010) Auxins reverse plant male sterility caused by high temperatures. Proc. Natl. Acad. Sci. USA 107: 8569–8574.

Salazar JD, Saithong T, Brown PE, Foreman J, Locke JC, Halliday KJ, Carre IA, Rand DA, Millar AJ (2009) Prediction of photoperiodic regulators from quantitative gene circuit models. Cell 139: 1170–1179.

Salome PA, McClung CR (2005) PSEUDO-RESPONSE REGULATOR 7 and 9 are partially redundant genes essential for the temperature responsiveness of the *Arabidopsis* circadian clock. Plant Cell 17: 791–803.

Salome PA, Weigel D, McClung CR (2010) The role of the *Arabidopsis* morning loop components CCA1, LHY, PRR7, and PRR9 in temperature compensation. Plant Cell 22: 3650–3661.

Searle I, He Y, Turck F, Vincent C, Fornara F, Kroeber, S, Amasino RA, Coupland G. (2006) The transcription factor FLC confers a flowering response to vernalization by repressing meristem competence and systemic signaling in *Arabidopsis*. Genes Dev. 20: 898–912.

Sellaro R, Pacin M, Casal JJ (2012) Diurnal dependence of growth responses to shade in *Arabidopsis*: Role of hormone, clock, and light signaling. Mol. Plant 5: 619–628.

Semenov M, Shewry P (2011) Modelling predicts that heat stress, not drought, will increase vulnerability of wheat in Europe. Sci. Rep. 1: 1–5.

Seo PJ, Park MJ, Lim MH, Kim SG, Lee M, Baldwin IT, Park CM (2012) A self-regulatory circuit of CIRCADIAN CLOCK-ASSOCIATED1 underlies the circadian clock regulation of temperature responses in *Arabidopsis*. Plant Cell 24: 2427–2442.

Sessa G, Carabelli M, Sassi M, Ciolfi A, Possenti M, Mittempergher F, Becker J, Morelli G, Ruberti I (2005) A dynamic balance between gene activation and repression regulates the shade avoidance response in *Arabidopsis*. Genes Dev. 19: 2811–2815.

Shen H, Zhu L, Castillon A, Majee M, Downie B, Huq E (2008) Light-induced phosphorylation and degradation of the negative regulator PHYTOCHROME-INTERACTING FACTOR1 from *Arabidopsis* depend upon its direct physical interactions with photoactivated phytochromes. Plant Cell 20: 1586–1602.

Shindo C, Aranzana MJ, Lister C, Baxter C, Nicholls C, Nordborg M, Dean C (2005) Role of FRIGIDA and FLOWERING LOCUS C in determining variation in flowering time of *Arabidopsis*. Plant Physiol. 138: 1163–1173.

Somers DE, Webb AA, Pearson M, Kay SA (1998) The short-period mutant, toc1-1, alters circadian clock regulation of multiple outputs throughout development in *Arabidopsis thaliana*. Development 125: 485–494.

Song YH, Lee I, Lee SY, Imaizumi T, Hong JC (2012a) CONSTANS and ASYMMETRIC LEAVES 1 complex is involved in the induction of FLOWERING LOCUS T in photoperiodic flowering in *Arabidopsis*. Plant J. 69: 332–342.

Song YH, Smith RW, To BJ, Millar AJ, Imaizumi T (2012b) FKF1 conveys timing information for CONSTANS stabilization in photoperiodic flowering. Science 336: 1045–1049.

Stavang JA, Gallego-Bartolome J, Gomez MD, Yoshida S, Asami T, Olsen JE, Garcia-Martinez JL, Alabadi D, Blazquez MA (2009) Hormonal regulation of temperature-induced growth in *Arabidopsis*. Plant J. 60: 589–601.

Stockinger EJ, Gilmour SJ, Thomashow MF (1997) *Arabidopsis thaliana* CBF1 encodes an AP2 domain-containing transcriptional activator that binds to the C-repeat/DRE, a cis-acting DNA regulatory element that stimulates transcription in response to low temperature and water deficit. Proc. Natl. Acad. Sci. USA 94: 1035–1040.

Storkey J (2004) Modelling seedling growth rates of 18 temperate arable weed species as a function of the environment and plant traits. Ann. Bot. 93: 681–689.

Suarez-Lopez P, Wheatley K, Robson F, Onouchi H, Valverde F, Coupland G (2001) CONSTANS mediates between the circadian clock and the control of flowering in *Arabidopsis*. Nature 410: 1116–1120.

Sugano S, Andronis C, Green RM, Wang ZY, Tobin EM (1998) Protein kinase CK2 interacts with and phosphorylates the *Arabidopsis* circadian clock-associated 1 protein. Proc. Natl. Acad. Sci. USA 95: 11020–11025.

Sun J, Qi L, Li Y, Chu J, Li C (2012a) PIF4-mediated activation of YUCCA8 expression integrates temperature into the auxin pathway in regulating *Arabidopsis* hypocotyl growth. PLoS Genet. 8: e1002594.

Sun X, Kang X, Ni M (2012b) Hypersensitive to red and blue 1 and its modification by protein phosphatase 7 are implicated in the control of *Arabidopsis* stomatal aperture. PLoS Genet. 8: e1002674.

Tepperman JM, Zhu T, Chang HS, Wang X, Quail PH (2001) Multiple transcription-factor genes are early targets of phytochrome A signaling. Proc. Natl. Acad. Sci. USA 98: 9437–9442.

Thines B, Harmon FG (2010) Ambient temperature response establishes ELF3 as a required component of the core *Arabidopsis* circadian clock. Proc. Natl. Acad. Sci. USA 107: 3257–3262.

Thomashow MF (1999) PLANT COLD ACCLIMATION: Freezing tolerance genes and regulatory mechanisms. Annu. Rev. Plant. Physiol. Plant. Mol. Biol. 50: 571–599.

Thomashow MF (2010) Molecular basis of plant cold acclimation: Insights gained from studying the CBF cold response pathway. Plant Physiol. 154: 571–577.

Thomashow MF (2001) So what's new in the field of plant cold acclimation? Lots! Plant Physiol. 125: 89–93.

Thuiller W, Lavorel S, Araujo MB, Sykes MT, Prentice IC (2005) Climate change threats to plant diversity in Europe. Proc. Natl. Acad. Sci. USA 102: 8245–8250.

Valverde F, Mouradov A, Soppe W, Ravenscroft D, Samach A, Coupland G (2004) Photoreceptor regulation of CONSTANS protein in photoperiodic flowering. Science 303: 1003–1006.

Vijayan P, Browse J (2002) Photoinhibition in mutants of *Arabidopsis* deficient in thylakoid unsaturation. Plant Physiol. 129: 876–885.

Vince-Prue D (1997) Photoperiodism in plants (2nd edn). Academic Press, London.

Vogel JT, Zarka DG, Van Buskirk HA, Fowler SG, Thomashow MF (2005) Roles of the CBF2 and ZAT12 transcription factors in configuring the low temperature transcriptome of *Arabidopsis*. Plant J. 41: 195–211.

Wilczek AM, Roe JL, Knapp MC, Cooper MD, Lopez-Gallego C, Martin LJ, Muir CD, Sim S, Walker A, Anderson J, Egan JF, Moyers BT, Petipas R, Giakountis A, Charbit E, Coupland G, Welch SM, Schmitt J (2009) Effects of genetic perturbation on seasonal life history plasticity. Science 323: 930–934.

Wilczek A, Burghardt L, Cobb A, Cooper M, Welch S, Schmitt J (2010) Genetic and physiological bases for phenological responses to current and predicted climates. Philos. Trans. R. Soc. Lond. B: Biol. Sci. 12: 3129.

Zhang XN, Wu Y, Tobias JW, Brunk BP, Deitzer GF, Liu D (2008) HFR1 is crucial for transcriptome regulation in the cryptochrome 1-mediated early response to blue light in *Arabidopsis thaliana*. PLoS One 3: e3563.

Zhang Y, Liu Z, Liu R, Hao H, Bi Y (2011a) Gibberellins negatively regulate low temperature-induced anthocyanin accumulation in a HY5/HYH-dependent manner. Plant Signal. Behav. 6: 632–634.

Zhang Y, Yang J, Guo S, Meng J, Zhang Y, Wan S, He Q, Li X (2011b) Over-expression of the *Arabidopsis* CBF1 gene improves resistance of tomato leaves to low temperature under low irradiance. Plant Biol. 13: 362–367.

Zhang Y, Zheng S, Liu Z, Wang L, Bi Y. (2011c) Both HY5 and HYH are necessary regulators for low temperature-induced anthocyanin accumulation in *Arabidopsis* seedlings. J. Plant Physiol. 168: 367–374.

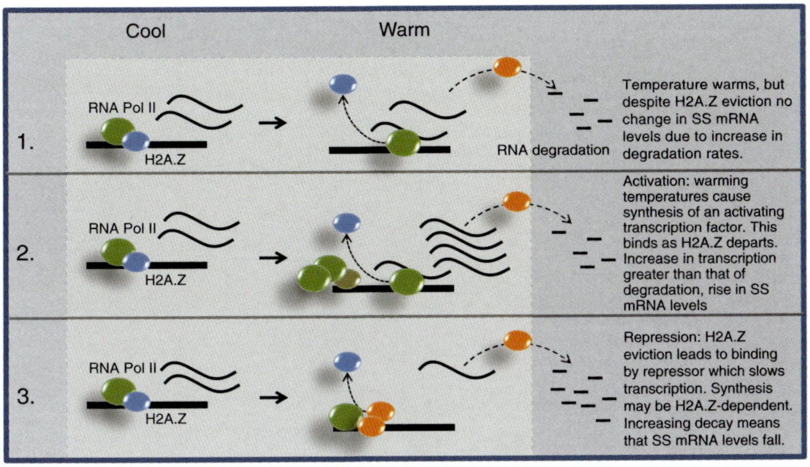

Figure 1.2 Control of SS transcript abundance by the temperature-dependent association of H2A.Z-containing nucleosomes with transcription start sites. (1) A passive affect alone does not produce an increase in SS levels, because mRNA degradation rates too are affected. (2) The eviction of H2A.Z allows an activator to bind. The abundance of this may also be temperaturecontrolled or a positive feedback through autoactivation. (3) Eviction allows a repressor to bind, allowing nucleosome eviction to depress transcription rates. After Kumar and Wigge (2010) and Franklin (2010). RNA Pol II– RNA polymerase II.

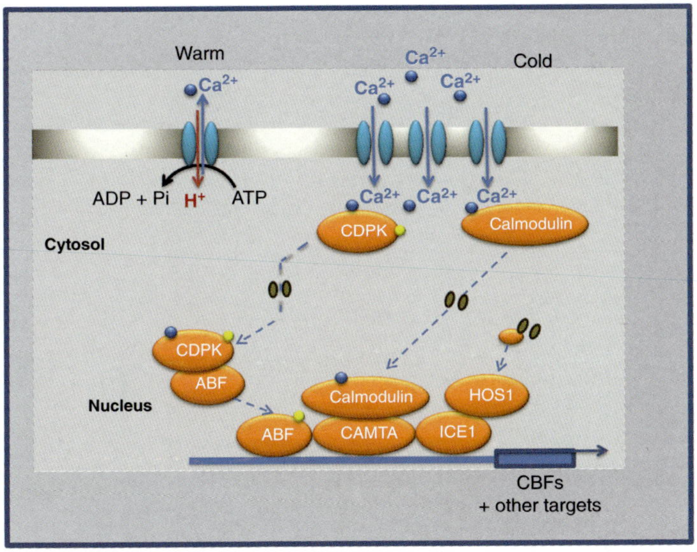

Figure 1.3 Schematic of our current understanding of cold perception and signaling in plants, showing possible signal transduction pathways from cytosolic free calcium increases to transcriptional control. For abbreviations see main text.

Temperature and Plant Development, First Edition. Edited by Keara A. Franklin and Philip A. Wigge.

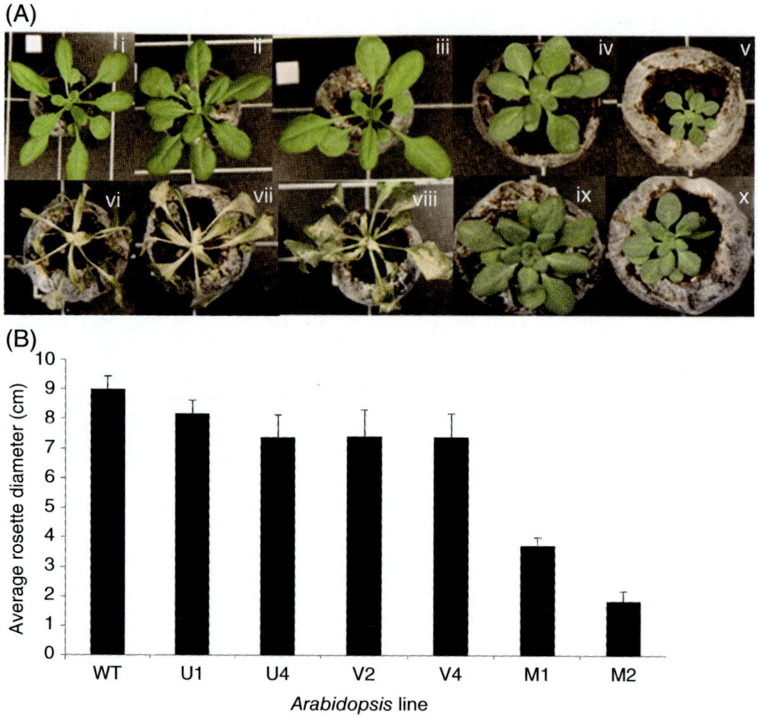

Figure 2.1 Effect of *Vaccinium* CBF overexpression upon freezing tolerance and development in transgenic *Arabidopsis*. (A) Photographs show transgenic *Arabidopsis* expressing *Vaccinium* CBF: *V. myrtillus* (iv, v, ix, x), *V. uliginosum* (ii, vii), *V. vitis-idaea* (iii, viii), and WT (i, vi), before (i–v) and after (vi–x) freezing (−7°C for 24 h in the dark). (B) Bar chart showing average rosette diameter of 'WT' *Arabidopsis* and lines overexpressing either *V. myrtillus*, *V. uliginosum*, or *V. vitis-idaea* CBF ('M', 'U', and 'V', respectively), $n = 8$, error bars are standard errors of the mean. Reproduced from Oakenfull *et al.* (2013).

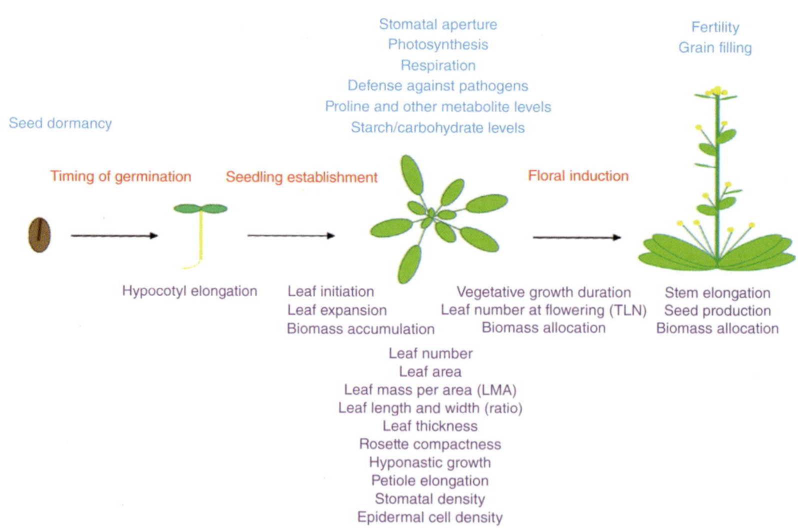

Figure 3.1 Schematic overview of morphological/developmental (*purple*), physiological (*blue*), and phase transition (*red*) traits affected by high temperature throughout the life history of the flowering rosette plant *A. thaliana*.

Figure 3.2 Morphological acclimation of *A. thaliana* to increased ambient temperature and high-temperature stress. (A, B) Rosette morphology of 30-day-old plants grown continuously at (A) 22°C or (B) 27°C, under LD conditions. Note the open rosette structure in warm-grown plants with elongated petioles, erect leaves, early flowering (elongated inflorescence), lower pigmentation (chlorophyll), and thin leaves (B) compared to those in panel (A). (C, D) Heat-induced hyponastic growth phenotype of (C) Col-0 plant grown at 20°C and (D) the same plant treated for 6h with high temperature (38°C). Note the orange-painted dots, used to facilitate leaf angle quantification. (E, F) High-temperature-induced hypocotyl elongation. Seven-day-old Col-0 seedling grown at (E) 22°C and (F) 27°C. Note that high temperature induces not only hypocotyl elongation but also hyponastic growth of cotyledons. Scale bar is 5mm. (G, H) Effect of DIF treatment on whole rosette morphology. (G) Plants grown under + DIF (22°C day, 12°C night) and (H) −DIF (12°C day, 22°C night). Note the typical compact rosette phenotype of plants grown under −DIF.

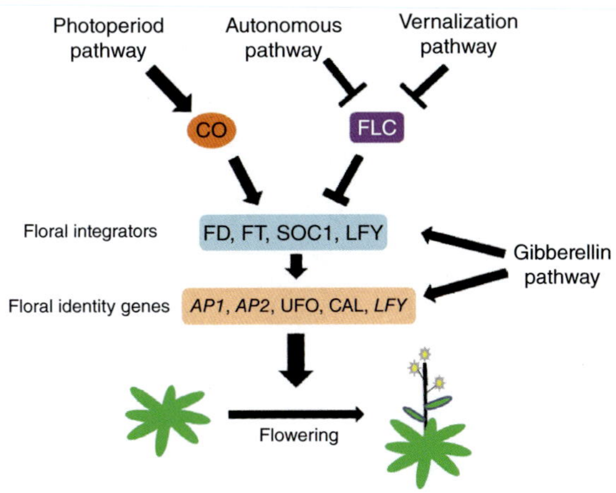

Figure 4.1 Four major flowering pathways in *Arabidopsis*. Vernalization and autonomous pathways commonly act to repress *FLC*. *FLC* acts to repress a group of floral integrators, including FT and SOC1, to block rapid flowering. The photoperiod pathway acts to activate floral integrators under inductive photoperiods. Activated floral integrators activate floral identity genes, such as *APETALA 1* (*AP1*), *APETALA 2* (*AP2*), *SEPALATA 3* (*SEP3*), *FRUITFUL* (*FUL*), and *LFY* to initiate flower development. GA-mediated pathway acts directly to downstream floral activators, including *SOC1* and *LFY*.

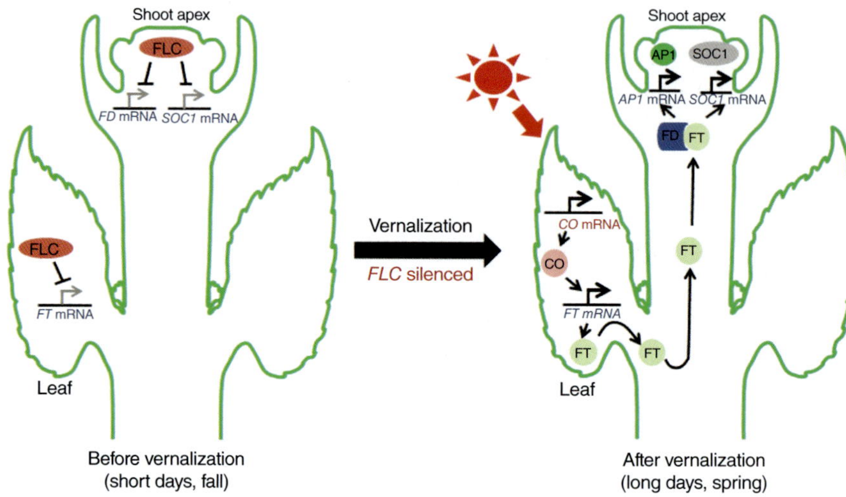

Figure 4.2 Vernalization-mediated flowering in *Arabidopsis*. Prior to vernalization, high levels of FLC prevent *FT*, *FD*, and *SOC1* expression and flowering is inhibited. After vernalization, *FLC* is silenced. In the absence of a prohibitory effect by FLC, floral integrators, FT, FD, and SOC1, can be activated by inductive photoperiods. FT protein has been shown to be a mobile floral activator which initiates flowering.

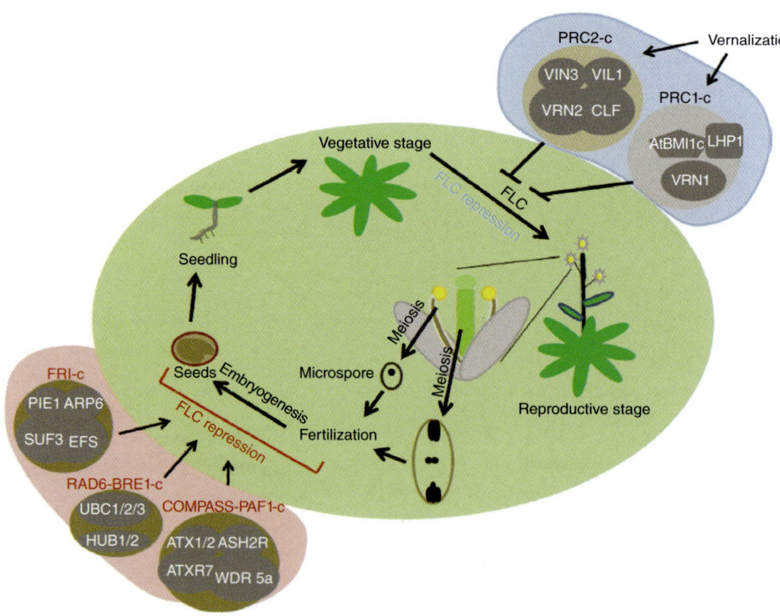

Figure 4.3 Fates of *FLC* chromatin during the *Arabidopsis* life cycle. Repressive complexes, PRC2 and PRC1, cause mitotically stable repression of *FLC* by vernalization. After meiosis, repressed *FLC* is reactivated by activating chromatin-remodeling complexes, including the FRIGIDA-containing complex (FRI-c), RAD6/BRE1-containing complex (RAD6-BRE1-c), and COMPASS/PAF1-containing complex (COMPASS-PAF1-c).

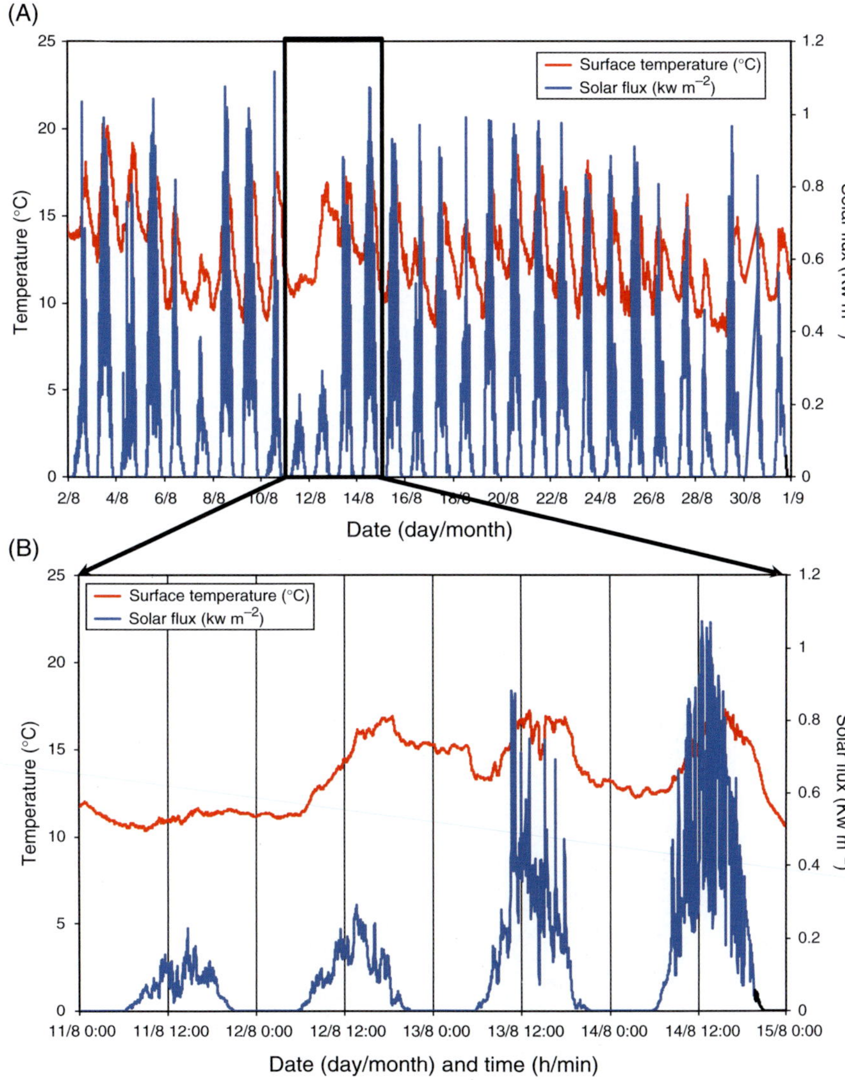

Figure 5.1 Variation in light and temperature provides a constantly changing environment for plants. There is an underlying pattern reflecting day and night, but a high degree of noise is superimposed over this; thus, a plant will have to be constantly monitoring and responding to environmental change. (A) Ground temperature and solar flux across the month of August 2011 in Edinburgh, UK. (B) Effect of rainy, cloudy days. Expanded panel showing the extent of day on day variation across consecutive days. Note the very small temperature range on the 11th of August and the rapid changes in solar flux on the 14th of August. (C) Relative humidity across August 2011. Weather data courtesy of Massimo Vieno, School of GeoSciences, Edinburgh University.

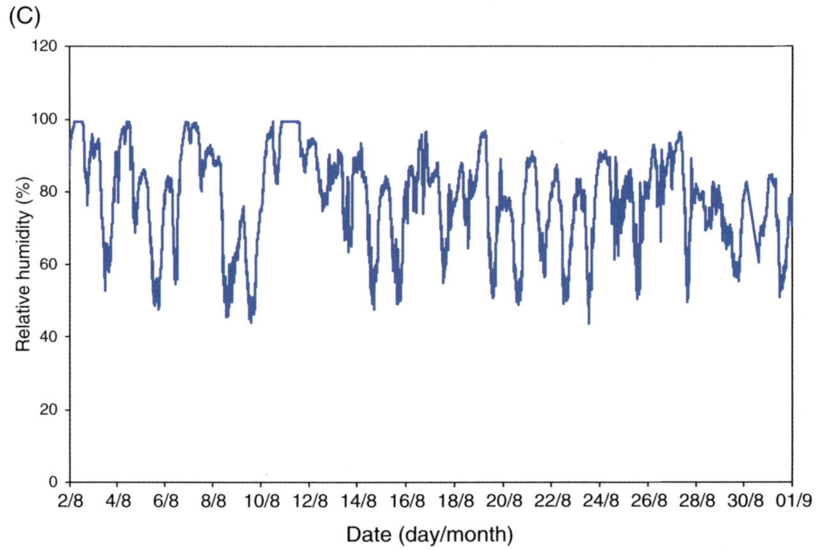

Figure 5.1 (*Continued*)

Figure 5.2 Phenotype of 46-day-old WT (Ler) on left and *phyABD* mutant *Arabidopsis* plants on right grown in white light, under short days (8 h light:16 h darkness) at 21°C.

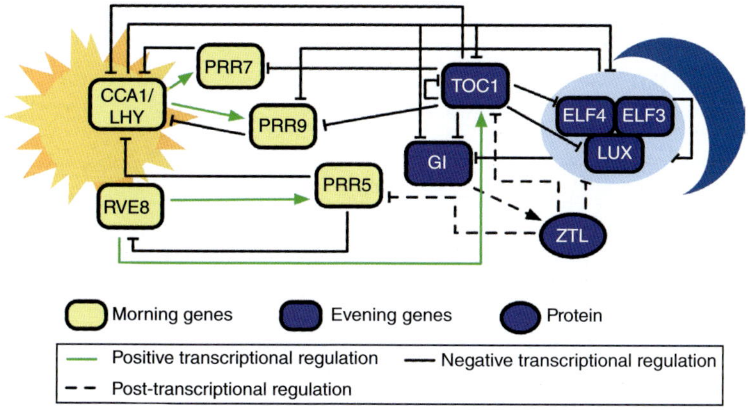

Figure 6.1 Model of the circadian clock in *Arabidopsis*. Due to the complexity of the circadian clock in *Arabidopsis*, a simplified version is shown to emphasize factors currently known to be involved in temperature responses. Transcriptional regulation is emphasized in this model and depicted by *solid* lines. *Green* and *black* lines represent positive and negative regulation, respectively. The *yellow* blocks are genes expressed during the day; *blue* blocks represent evening-expressed genes. *Dashed* lines indicate posttranscriptional regulation with ZTL, the only protein depicted in the model. The EC is represented by the *light blue* circle surrounding *ELF3*, *ELF4*, and *LUX*. Lines that meet the blue circle indicate regulation of the entire complex, whereas lines leading to specific blocks indicate regulation of that gene alone.

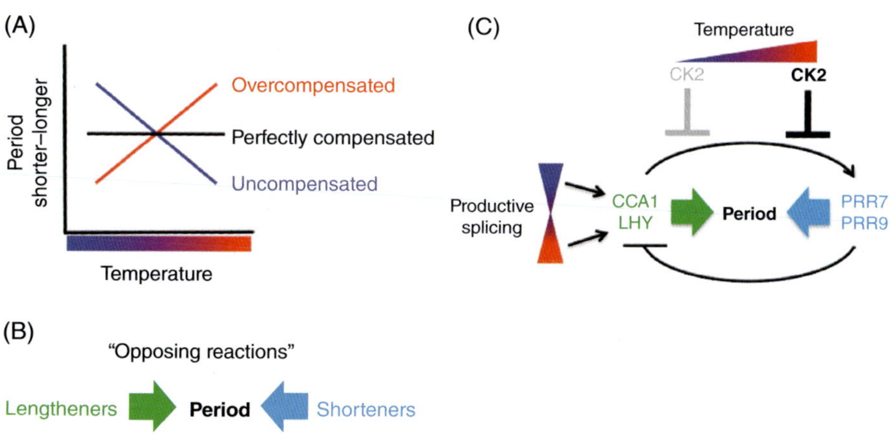

Figure 6.2 Temperature compensation relies on a balance of reactions. (A) Schema illustrating the period length associated with compensated (*black*), overcompensated (*red*), and uncompensated (*blue*) clocks over a range of temperatures. A wild-type clock that is perfectly compensated across all physiological temperatures maintains a constant period. (B) The 'Opposing Reactions' model of temperature compensation. The period is maintained relatively constant by the balance of period lengtheners and period shorteners. (C) An example of this balance is shown with CCA1/LHY and PRR7/PRR9 that are tightly regulated by several temperature-sensitive mechanisms such as CK2 phosphorylation and AS. The *cca1 lhy* double mutant has a short period suggesting CCA1 and LHY are period lengtheners, whereas the *prr7 prr9* double mutant has a long period suggesting they function as period shorteners. A balance of activities of these genes results in a compensated clock.

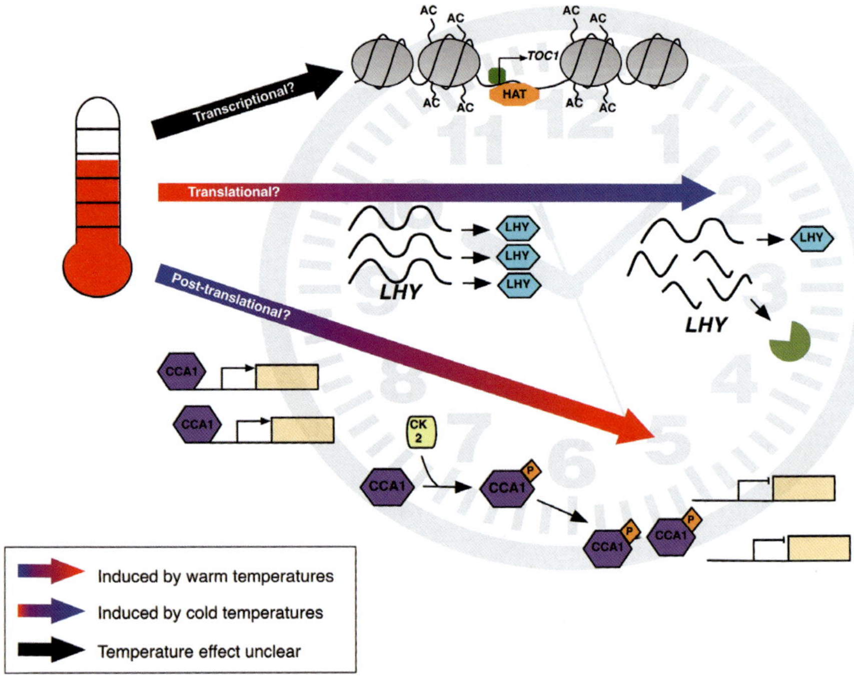

Figure 6.3 Possible points of temperature integration into the circadian clock. It is still not clear how the clock senses changes in ambient temperature. The three processes depicted in this figure integrate temperature at the level of transcriptional, translational, and posttranslational regulation. The *black arrow* represents the transcriptional regulation of chromatin remodeling that has been shown to be important for *TOC1* regulation but not yet linked to temperature. However, high temperature has been shown to deplete H2A.Z nucleosome occupancy leading to activation of the warm temperature transcriptome (Kumar and Wigge 2010). The translational arrow describes the temperature-dependent AS events for *LHY*. As the temperature cools (indicated by a transition from *red* to *blue* in the *arrow*), there is an increase in unproductive splice forms for *LHY* that are likely responsible for the decrease in protein abundance. Finally, the posttranslational arrow depicts the high-temperature-dependent (illustrated by a *blue* to *red* color transition) phosphorylation of CCA1 by CK2, leading to a decrease in CCA1 activity. The exact timing or sensitivity of this response is unclear but may be an important mechanism for fine-tuning the temperature response.

6 Temperature and the circadian clock

Kathleen Greenham and C. Robertson McClung

6.1 Introduction

Plants are remarkably adaptable to ecological stresses such as pathogen attack, shading from a nearby plant, drought, or temperature. Each of these stresses elicits a cascade of signaling events within the plant, resulting in physiological responses naturally selected to maximize growth and fitness. These signaling events are coordinated with the daily light–dark and warm–cold cycles present on Earth. Perhaps one of the most startling examples of plant preparedness is the anticipation of tomorrow. For example, a sunflower plant will reposition its leaves at the end of the night to face the east in anticipation of dawn to exploit light absorption at the onset of dawn (Shell and Lang 1976) (http://plantsinmotion.bio.indiana.edu/plantmotion/movements/tropism/solartrack/solartrack.html). This remarkable ability to anticipate the arrival of sunlight and subsequent warming is dependent on an internal biological clock. This biological clock is essential for synchronizing diurnal environmental changes such as temperature, a major signal affecting clock function (McClung and Davis 2010). This chapter will focus on addressing the current understanding of the relationship between temperature and the circadian clock in plants.

The first experiments describing circadian rhythms were reported in 1729 by the French astronomer de Mairan who observed that plant leaves exhibit daily rhythmic movements that persist in constant darkness (de Mairan 1729). This innate diurnal periodicity was measured a century later by de Candolle and others and found to be approximately, but not exactly, 24 h in length (de Candolle 1832). These astute observations revealed an internal mechanism within the plant that was entrained to the 24 h day and night cycle present on Earth. The term circadian, coined in 1959 (Halberg et al. 1959) from the Latin words 'circa' (about) and 'dies' (day), is used to describe the 24 h period that signifies one complete cycle of these biological rhythms.

Temperature and Plant Development, First Edition. Edited by Keara A. Franklin and Philip A. Wigge.

An endogenous circadian clock is present in all three domains of life, eubacteria, archaea, and eukaryotes. This includes cyanobacteria, halobacteria, fungi, plants, invertebrates, and vertebrates such as mice and humans (Bell-Pedersen et al. 2005; Foster and Roenneberg 2008; Whitehead et al. 2009; Edgar et al. 2012). As humans, the most noticeable workings of our internal clock are our sleep patterns. The desire to sleep when it is dark and be awake when it is light is one manifestation of the orchestrated entrainment of our metabolism, physiology, and behavior to the Earth's rotation (Sahar and Sassone-Corsi 2012). Although the molecular components of the circadian clock across biological taxa are quite different (Bell-Pedersen et al. 2005; Zhang and Kay 2010), the defining characteristics of circadian rhythms are conserved. Circadian rhythms display three fundamental properties: self-sustaining periodicity, entrainment, and temperature compensation (Harmer et al. 2001). Self-sustaining periodicity refers to the persistence of a circadian rhythm under constant conditions, first demonstrated by de Mairan for leaf movement in constant darkness (DD). A true circadian rhythm can be entrained to environmental cues, termed *zeitgeber* ('time giver' in German). The strongest *zeitgeber* is the light/dark (LD) cycle, although temperature cycles also are effective *zeitgebers* (Harmer et al. 2001). The final property of circadian rhythms, and one focus of this chapter, is temperature compensation, whereby the period remains relatively constant at varying physiologically relevant temperatures.

The widespread occurrence of an endogenous circadian oscillator across diverse taxa suggests that there is a fitness advantage in entraining internal processes to the external environment. In plants, the circadian clock is involved in a vast array of processes, from stomatal conductance and photosynthesis to flowering time, stem elongation, leaf movement, resistance to microbial pathogen invasion, and even herbivory resistance (Yakir et al. 2007; Resco et al. 2009; Goodspeed et al. 2012). The importance of the clock in plants is evident in the breadth of circadian regulation of gene expression. Transcriptomic analyses have shown that genes known to be involved in most, if not all, facets of plant growth are circadian regulated (Doherty and Kay 2010). These studies predict between 6% and 90% of the transcriptome in the model plant *Arabidopsis thaliana* is circadian regulated, depending on the condition examined (Harmer 2009; Doherty and Kay 2010). Evidently, the environmental conditions and the methods used to assess the level of transcriptomic regulation by the clock provide different estimates; however, there is no question as to the involvement of the clock in transcript accumulation. In *Arabidopsis*, proper clock regulation of growth confers a fitness advantage compared to long- and short-period mutants whose endogenous periods do not resonate with the environmental cycle (Dodd et al. 2005b; Resco et al. 2009; Yerushalmi and Green 2009).

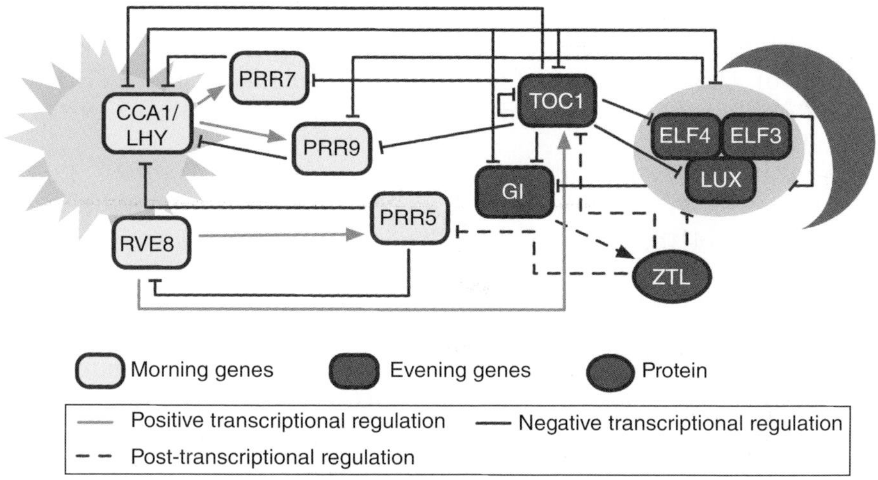

Figure 6.1 Model of the circadian clock in *Arabidopsis*. Due to the complexity of the circadian clock in *Arabidopsis*, a simplified version is shown to emphasize factors currently known to be involved in temperature responses. Transcriptional regulation is emphasized in this model and depicted by *solid* lines. *Green* and *black* lines represent positive and negative regulation, respectively. The *yellow* blocks are genes expressed during the day; *blue* blocks represent evening-expressed genes. *Dashed* lines indicate posttranscriptional regulation with ZTL, the only protein depicted in the model. The EC is represented by the *light blue* circle surrounding *ELF3*, *ELF4*, and *LUX*. Lines that meet the blue circle indicate regulation of the entire complex, whereas lines leading to specific blocks indicate regulation of that gene alone. For color detail, please see color plate section.

This competitive advantage has also been demonstrated in cyanobacteria where mutant strains with altered circadian period outcompete wild-type strains when grown in a culture where the LD cycle length matches the period of the mutant (Ouyang et al. 1998; Woelfle et al. 2004).

A common feature of circadian oscillators in all eukaryotes is the presence of negative transcriptional feedback loops. The circadian clock in plants is comprised of a series of interlocked regulatory feedback loops that can be simplistically divided into a central loop coupled to a morning and an evening loop, defined by the timing of expression of their components (Figure 6.1). The 'morning loop' consists of two MYB transcription factors, CIRCADIAN CLOCK-ASSOCIATED 1 (CCA1) and LATE ELONGATED HYPOCOTYL (LHY), whose expression peaks at dawn (Schaffer et al. 1998; Wang and Tobin 1998). The arrhythmic phenotypes caused by overexpression of CCA1 or LHY were important indicators that these proteins are vital components of the clock (Barak et al. 2000). Both CCA1 and LHY activate the expression of two additional morning loop components, PSEUDO-RESPONSE REGULATOR (PRR) proteins PRR7 and PRR9 (Farré et al. 2005; Nakamichi et al. 2005, 2010). PRR9,

PRR7, and another member of the family, PRR5, complete the loop by repressing the expression of *LHY* and *CCA1* (Harmer 2009; Nakamichi et al. 2010). The expression of *PRR5* is positively regulated by REVEILLE8 (RVE8), an myb-like transcription factor closely related to CCA1/LHY (Rawat et al. 2011).

The PRR family is made up of five members, PRR9, PRR7, PRR5, PRR3, and PRR1/TIMING OF *CAB* EXPRESSION1 (TOC1). TOC1 is a component of the 'evening loop' along with the evening complex (EC) that includes EARLY FLOWERING 3 (ELF3), ELF4, and LUX ARRHYTHMO (LUX) (Helfer et al. 2011; Nusinow et al. 2011). The EC is a recent addition to the evening loop even though the necessity of each member for maintaining plant circadian rhythms has been known for some time (Hicks et al. 1996a; Doyle et al. 2002; Hazen et al. 2005; Kolmos et al. 2009). Although LUX is the only member of the complex shown to bind DNA directly, both ELF3 and ELF4 are important for proper expression of the EC transcriptional targets (Kolmos et al. 2009; Helfer et al. 2011; Nusinow et al. 2011; Herrero et al. 2012).

Apart from transcriptional circuitry, there is substantial posttranslational regulation in the clock network. Both TOC1 and PRR5 protein levels are regulated through an SCF-dependent degradation mediated by the F-box protein ZEITLUPE (ZTL) (Harmer 2009; Somers and Fujiwara 2009). ZTL protein stabilization is dependent on another protein, GIGANTEA (GI) (Somers et al. 2000; Han et al. 2004; Kim et al. 2007). GI was originally identified by the extremely late-flowering phenotype of several *gi* mutant alleles compared to wild type (Rédei 1962; Koornneef et al. 1991), although subsequent studies revealed that GI has an important role in the circadian oscillator as well (Fowler et al. 1999; Park et al. 1999). The *GI* transcript shows rhythmic expression under (LL) light conditions, and this pattern is disrupted in genotypes with defective circadian clocks, such as *CCA1* and *LHY* overexpression lines (Fowler et al. 1999). The stabilizing effect of GI on ZTL confers a circadian rhythm in ZTL protein abundance, even though ZTL transcript does not oscillate. In addition, the release of accumulated ZTL from GI at dusk causes a rapid increase in the rate of TOC1 and PRR5 degradation (Kim et al. 2007).

There are several layers of negative regulation between the morning and evening loops. The regulation of *CCA1/LHY* by TOC1 is supported by abundant genetic data, including altered expression of *CCA1*, *LHY*, and *GI* following TOC1 overexpression (Pruneda-Paz and Kay 2010; Somers 2012). Characterization of the TOC1 protein has revealed DNA-binding capability and transcriptional-repression activity (Gendron et al. 2012; Huang et al. 2012; Pokhilko et al. 2012; Somers 2012). TOC1, as shown by chromatin immunoprecipitation, binds to the promoters and represses the expression of the morning elements *CCA1*, *LHY*, *PRR9*, and *PRR7*, as well as members of the evening loop, *GI*, *TOC1*, *ELF4*, and *LUX* (Huang et al. 2012), revealing extensive feedback regulation within and between the core loops.

The EC also regulates the morning loop by repressing *PRR9* as well as negatively regulating its own expression (Kolmos et al. 2009; Dixon et al. 2011; Helfer et al. 2011; Chow et al. 2012). Similarly, the morning loop interacts transcriptionally with the evening loop through the negative regulation of CCA1/LHY on *TOC1* (Alabadí et al. 2001; Ding et al. 2007), *LUX* (Hazen et al. 2005), *ELF4* (Kikis et al. 2005; Li et al. 2011), and *GI* (Locke et al. 2005). *TOC1* expression is also regulated through an H3 acetylation-dependent process that is antagonized by CCA1. H3 acetylation at the Evening Element (EE) motif in the *TOC1* promoter leads to a relaxed chromatin structure, allowing activation of transcription (Perales and Más 2007; Más 2008). Figure 6.1 depicts a slightly modified version of a recent model of the clock network in *Arabidopsis* (Pokhilko et al. 2012); although this model incorporates a large amount of data, it is simplified for clarity.

The model shown in Figure 6.1 highlights the complexity of the transcriptional and posttranscriptional regulation of the circadian clock genes. This complex network is important for integrating signals from environmental and physiological cues. As mentioned at the beginning of this chapter, the clock controls important physiological, metabolic, and behavioral processes. The clock functions as a central integrator by relaying environmental signals to the necessary output pathways for a coordinated growth response. A good example of this clock-dependent coordination is the regulation of hypocotyl growth. Light, hormones, and temperature are some of the pathways that regulate hypocotyl elongation (Vandenbussche et al. 2005; Nozue and Maloof 2006). Many clock mutants show defects in hypocotyl elongation at least in part due to the inability to properly gate the growth period (Zagotta et al. 1996; Schaffer et al. 1998; Más et al. 2003; Nozue and Maloof 2006). Under LD conditions hypocotyl elongation occurs toward the end of the night cycle, a growth period that is defined by the EC (Nozue et al. 2007; Niwa et al. 2009; Nusinow et al. 2011). Therefore, the clock acts as a point of integration of these various signals to properly gate the hypocotyl growth response.

Another important environmental signal in the clock mechanism is temperature signaling, as marked by the inherent properties of temperature compensation and entrainment of the clock. A tremendous amount of work has led to the comprehensive model of the clock we have today; however, we still know very little about how the plant senses changes in temperature in terms of either daily fluctuations or seasonal effects (McClung and Davis 2010).

6.2 Temperature compensation

The fourth assessment report released in 2007 by the Intergovernmental Panel on Climate Change (IPCC) projects that low-latitude regions, where most global cereal production occurs, are at risk for yield reduction with a temperature increase of as little as +1°C and almost certainly cannot tolerate more than a +3°C rise before yields decline (Easterling et al. 2007). In view of these predictions, it is sobering that model simulations project worldwide temperature increases of 4–5°C by 2080 (Easterling et al. 2007). Improvement of predictions of the effect of temperature increase on crop yields requires better models to explain plant responses to temperature change. This will also depend on our ability to improve our experimental design to simulate the natural environment. Recent work suggests that current experimental data underpredicts the effects of warming on plant phenology in terms of spring leaf emergence and flowering time (Wolkovich et al. 2012). Plants are naturally equipped with mechanisms to respond to fluctuations in ambient temperature, as detailed elsewhere in this volume (Chapters 2 and 4), but we believe that the complex relationships between temperature and other environmental factors remain ill defined.

The importance of the circadian clock for fitness requires that it be reliable and accurate. This presented a particular puzzle for plants and poikilothermic animals whose internal temperatures vary in response to the environment, because a temperature-dependent clock would inevitably be inaccurate. Pittendrigh (1954) articulated the dilemma and performed a set of classic experiments with the *Drosophila pseudoobscura* where he measured eclosion rhythms and showed that the period was maintained at 24 h over a range of temperatures, from 16°C to 26°C. A similar lack of temperature sensitivity was observed for the phototactic response of the single-celled green alga *Euglena gracilis* for temperatures ranging from 16°C to 33°C (Bruce and Pittendrigh 1956). These early 'compensation' experiments led to the hypothesis that the clock must be temperature 'independent'. However, Hastings and Sweeney (1957) noticed that although the luminescence rhythm in the dinoflagellate *Gonyaulax polyedra* (recently renamed *Lingulodinium polyedrum*) was maintained with an average of 25.5 h, the length of the period was slightly longer at higher temperatures and shorter at lower temperatures. They concluded that, rather than independent of temperature, the clock was compensated (buffered) against temperature changes. The temperature coefficient or Q_{10} is a measure of the rate of change of a system following a 10°C increase in temperature. For typical biochemical reactions, this ranges from 2–3 (Ruoff 1992). However, the Q_{10} for circadian period was much less than 2, indicating compensation. For the *Gonyaulax* luminescence rhythm, Q_{10} was less than 1.0, suggesting 'overcompensation' (Figure 6.2A)

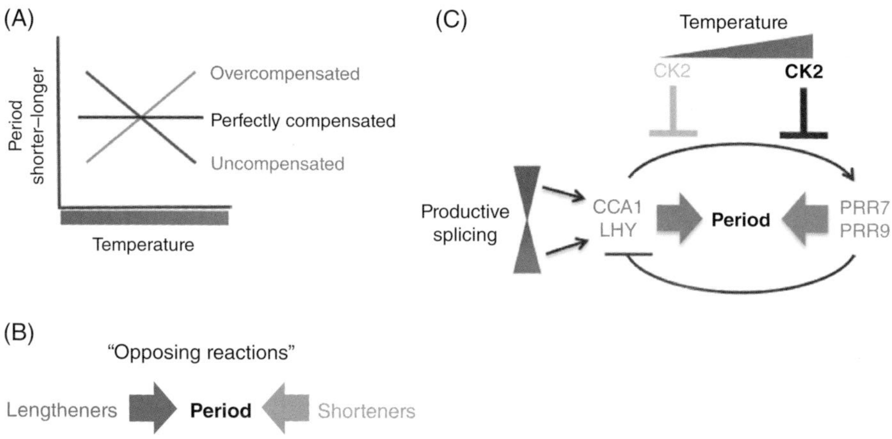

Figure 6.2 Temperature compensation relies on a balance of reactions. (A) Schema illustrating the period length associated with compensated (*black*), overcompensated (*red*), and uncompensated (*blue*) clocks over a range of temperatures. A wild-type clock that is perfectly compensated across all physiological temperatures maintains a constant period. (B) The 'Opposing Reactions' model of temperature compensation. The period is maintained relatively constant by the balance of period lengtheners and period shorteners. (C) An example of this balance is shown with CCA1/LHY and PRR7/PRR9 that are tightly regulated by several temperature-sensitive mechanisms such as CK2 phosphorylation and AS. The *cca1 lhy* double mutant has a short period suggesting CCA1 and LHY are period lengtheners, whereas the *prr7 prr9* double mutant has a long period suggesting they function as period shorteners. A balance of activities of these genes results in a compensated clock. For color detail, please see color plate section.

(Hastings and Sweeney 1957). This observation suggested that an active temperature-compensation mechanism was a natural feature of the endogenous biological clock.

One of the most confounding aspects of temperature compensation is how the temperature dependence of kinetic-rate constants, as defined by the Arrhenius equation (Bodenstein et al. 2011), can be reconciled with this clock property. A view reflecting a balancing of reactions with multiple feedback processes quickly became the model for interpreting the temperature effect on the clock (Hastings and Sweeney 1957; Karlsson and Johnsson 1972; Hoskins and Moore 2012). Perhaps one of the more detailed models described by Njus et al. (1974) postulated that ion oscillations and the regulation of transmembrane ion flux accounted for the feedback control necessary to explain the light and temperature responses. According to this model, changes in membrane lipid composition take several hours, thereby accounting for the delay in clock compensation following a temperature pulse (Njus et al. 1974). Plasma membrane viscosity alters with changes in temperature (Martinière et al.

2011), confirming a link between temperature and lipid composition. In addition, Ca^{2+} ion oscillations serve as a circadian-regulated signal that drives important processes in the plant (McAinsh et al. 1995; Trewavas 1999; Wood et al. 2001; Love et al. 2004). However, it seems that additional regulatory mechanisms have been recruited to implement temperature compensation.

Quantitative Trait Locus (QTL) studies of recombinant inbred line (RIL) populations derived from parents with similar clock characteristics under normal temperature conditions reveal transgressive segregation of circadian period and phase, meaning greater variation among the RILs than seen in the RIL parents (Swarup et al. 1999; Michael et al. 2003b). Transgressive inheritance suggests that the trait in question is controlled by additive effects of alleles at multiple loci. The architecture of the clock as multiple interlocked feedback loops is consistent with this interpretation. Loss of function of most clock genes alters period length rather than confers arrhythmia (Harmer 2009); consequently most clock genes can be thought of as period lengtheners (loss of function shortens period) and shorteners (loss of function lengthens period) (Figure 6.2B).

Similar logic can be applied to the analysis of temperature compensation. Natural variation studies in *Arabidopsis* suggest a complex and diverse genetic basis to temperature compensation. QTL analysis of several RIL populations in *Arabidopsis* suggests that multiple loci acting in *trans* are involved in regulating the levels of clock proteins to balance the response to changes in temperature (Edwards et al. 2005). Identifying the genes contributing to these *trans*-QTLs is a challenge using classical QTL mapping techniques and often requires additional analyses. One of the many *Arabidopsis* QTLs for period length that mapped only at 27°C was identified as *FLOWERING LOCUS C* (*FLC*) (Edwards et al. 2005), which encodes a MADS box transcription factor that inhibits flowering (Michaels and Amasino 1999; Sheldon et al. 1999). FLC has been shown to lengthen period in a dose-dependent manner, along with several other genes in the flowering time pathway (Salathia et al. 2006). Further transcriptomic and expression analysis implicated LUX as a candidate for mediating the effect of *FLC* on the clock at higher temperature (Edwards et al. 2005, 2006). It is likely that additional genes are contributing to this effect as predicted by the *trans*-QTLs.

A similar natural variation study was performed in *Brassica rapa* where 50 natural accessions and an RIL population of 159 individuals were screened for period length, phase, and amplitude at three temperatures. A wide variation in period length was observed with an inverse relationship between Q_{10} and period length, although Q_{10} values fell between 0.8 and 1.2, demonstrating temperature compensation in *B. rapa*. In addition, variation in period length was more extreme in the RILs compared to the accessions, suggesting that these period extremes are selected against in natural populations (Lou et al. 2011). These results further support the connection between a synchronized clock and fitness.

To uncover the mechanistic basis for temperature compensation, a number of studies have taken a genetic approach to identify mutants that are affected in this property. The discovery of mutants affected in period length led to a deeper understanding of the link between temperature compensation and the circadian clock. Several long-period mutants in *Neurospora* and *Drosophila* show temperature-compensation defects; some maintain temperature compensation but across a more narrow range of temperatures than the wild type; other mutants were found to have fully lost temperature compensation (Gardner and Feldman 1981; Huang et al. 1995; Matsumoto et al. 1999).

In *Arabidopsis*, genetic studies have focused on the role of the clock genes in the temperature response of the circadian clock. Gould et al. (2006) showed that the transcript accumulation levels of *TOC1*, *LHY*, and *CCA1* are all temperature dependent. For example, the amplitude and peak of *TOC1* expression increased at high temperature, while *LHY* showed the reciprocal response. At lower temperatures of 12°C, *CCA1*, *LHY*, and *GI* decreased in peak expression levels (Gould et al. 2006). Salomé et al. (2010) showed that *LHY* and *CCA1* transcript levels are equally affected by temperature. At 12°C, *CCA1*, *LHY*, and *TOC1* show a phase shift, peaking later than at 22°C. At 30°C, *CCA1*, *LHY*, and *TOC1* have weak expression without a clear peak. This was surprising considering wild-type plants are fairly well compensated at this temperature (Salomé et al. 2010). Consistent with Gould et al. (2006), *CCA1* and *LHY* show low levels of expression at elevated temperature (Salomé et al. 2010). Salomé et al. (2010) observed two peaks in *TOC1* transcript accumulation at 30°C, and the level of *TOC1* expression did not exceed the peak expression level at 22°C; however, *TOC1* transcript remained relatively high at all time points and did not reach the trough levels observed at 22°C. From these studies, it is evident that there are temperature effects on the expression of these clock genes, but further studies are required to clarify these discrepancies. As discussed later in this chapter, alternative splicing (AS) also contributes to temperature-dependent transcript regulation (James et al. 2012). It may prove to be more informative to examine the temperature effects on protein levels in order to interpret the functional implications of these changes (Greenbaum et al. 2003).

In addition to *CCA1*, *LHY*, and *TOC1*, there is a clear involvement of *GI* in the temperature-compensation mechanism (Gould et al. 2006). The exact role of *GI* in this context is uncertain due to the complexity of accumulated genetic data (Rédei 1962; Huq et al. 2000; Mizoguchi et al. 2005; Gould et al. 2006; Martin-Tryon et al. 2007; Sawa et al. 2007; Sawa and Kay 2011). Genetic screens over the years have identified a series of *gi* alleles showing a range of developmental phenotypes. As mentioned earlier, the first alleles were identified due to their severe late-flowering phenotypes (Rédei 1962), and many additional alleles have been identified on the basis of late flowering (Mizoguchi et al. 2005; Kim et al. 2007; Sawa et al. 2007; Sawa and Kay 2011). New *gi* alleles emerged with photomorphogenic phenotypes under red and blue light, implicating GI in light signaling (Huq et al. 2000; Martin-Tryon et al. 2007).

In addition to defects in flowering time and response to light, the available *gi* mutants exhibit altered temperature compensation. A line homozygous for one presumptive null allele, *gi-11*, had a period length at 17°C that matched with wild type but shortened relative to wild type when transferred to 12, 22, or 27°C (Gould et al. 2006). A line homozygous for a second presumptive null allele, *gi-201*, showed a slightly different response, with wild-type period lengths at both 22°C and 30°C and a slightly shorter period at 16°C (Salomé et al. 2010). This study also tested a line homozygous for a third mutant allele, *gi-1*, that has a 5-bp deletion leading to a premature stop codon, causing a loss of 171 amino acids from the C-terminus. This *gi-1* mutant had the same period as *gi-201* at 16°C but became increasingly shorter at higher temperatures, showing a much stronger phenotype than the presumptive null alleles (Salomé et al. 2010). The discrepancies between these two studies might reflect differences among alleles but might also be attributed to the different output rhythms monitored for assaying period length. Gould et al. (2006) measured cotyledon movement and therefore likely measured clock activity in the petiole, whereas Salomé et al. (2010) measured the luciferase activity of both *CCA1pro::Luc* and *TOC1pro::Luc* reporters in the cotyledons (e.g., Salomé et al. 2008). This may suggest tissue-specific clocks that are buffered differently to environmental signals. For example, the circadian clock in root tissue shows a different oscillation pattern than the clock in the shoot (Thain et al. 2000; James et al. 2008). In addition, certain clock output genes have different levels of temperature sensitivity leading to differential phase shifting, suggesting regulation by multiple circadian oscillators (Michael et al. 2003a).

A common trend among several mutants showing altered temperature response is conditional temperature sensitivity. Several *gi* mutants show normal period lengths at low temperature but become sensitive to high temperature. A similar trend is observed with members of the *PRR* family. The *prr7 prr9* mutant showed a wild-type period length at 12°C, but period lengthened at higher temperatures, demonstrating overcompensation (Salomé et al. 2010). In contrast, the *prr5 prr7* mutant had a slightly shorter period than wild type at 17°C and 22°C and became even shorter at 30°C.

These two observations suggest that members of the *PRR* family contribute differentially to the temperature response (Salomé et al. 2010). In addition, RVE8, involved in a negative feedback loop with PRR5, also contributes to the temperature response. Overexpression of RVE8 resulted in a short period at 17°C, 22°C, and 27°C but not at 12°C, while the *rve8-1* null mutant showed longer periods at 22°C and 27°C, but not 12°C or 17°C (Rawat et al. 2011). This suggested that PRR5 and RVE8 are important for the high-temperature-compensation response. *LHY*, *TOC1*, and *ZTL* loss-of-function mutants also showed increased sensitivity to changes in temperature; *lhy-20* and *toc1-101* are short-period mutants and ran shorter at 30°C compared to 12°C, while the long-period *ztl-4* mutant lengthened period at lower temperature (Salomé et al. 2010). These studies suggested that different compensatory mechanisms might be involved in regulating the high- *versus* low-temperature response.

The morning loop components CCA1, LHY, PRR7, and PRR9 play an important role in the temperature-compensation mechanism (Figure 6.2C). As mentioned, high temperature causes period lengthening in the *prr7 prr9* mutant. Temperature compensation is restored following an miRNA-induced reduction of *CCA1* and *LHY* expression, demonstrating that the role of PRR7 and PRR9 at high temperature is completely dependent on *CCA1* and *LHY* (Salomé et al. 2010). In addition, reduction of *CCA1* and *LHY* expression in wild type and *prr7 prr9* resulted in a short-period phenotype at all temperatures, suggesting that the role of PRR7 and PRR9 in temperature compensation is the regulation of *CCA1* and *LHY* (Salomé et al. 2010). This is consistent with PRR7- and PRR9-dependent repression of *CCA1* and *LHY* (Farré et al. 2005; Nakamichi et al. 2010). CCA1 and LHY complete the feedback loop by positively regulating *PRR7* and *PRR9* expression (Farré et al. 2005).

The binding affinity of CCA1 to the promoters of both *PRR7* and *PRR9* increased at higher temperatures, and this was inhibited by the protein kinase CK2-dependent phosphorylation of CCA1 (Sugano et al. 1998; Daniel et al. 2004; Portolés and Más 2010). CK2 activity increased at high temperature and caused accumulation of phosphorylated CCA1 and possibly a reduction in CCA1 binding to *PRR7*, *PRR9*, *TOC1*, and *LUX* promoters. The reduction in activity of CCA1 would explain the observed misexpression of *TOC1pro::Luc* transcription at high temperature (Gould et al. 2006; Portolés and Más 2010). This mechanism is reminiscent of the CK2-dependent phosphorylation of the FREQUENCY (FRQ) clock protein in *Neurospora*. Unlike CCA1, the phosphorylation of FRQ led to its degradation, but the involvement of CK2 at high temperature was consistent in both systems (Mehra et al. 2009). However, it should be noted that Portolés and Más (2010) performed promoter binding and phosphorylation assays using overexpression versions of CCA1 and CKB4, the regulatory subunit of CK2. The potential for overexpression to introduce artifactual protein–protein interactions makes it problematic to use these results to extrapolate to the relationship between CCA1 and CKB4 circadian-regulated accumulation and temperature in wild-type plants.

CCA1 protein was shown to accumulate just before dawn and reached peak levels 1 h after dawn (Wang and Tobin 1998). Due to the circadian regulation of each of these proteins, it is difficult to predict the importance of the CKB4-CK2-dependent phosphorylation of CCA1 at high temperature without detailed information regarding CKB4-CK2 protein levels. CKB4 protein levels were found to be circadian regulated and began to accumulate near the end of the day (Perales et al. 2006). At normal temperatures, phosphorylation of CCA1 by CK2 is required for formation of CCA1–DNA complexes (Daniel et al. 2004), suggesting that the phospho-dependent regulation of CCA1 is pervasive. This was the case for FRQ in *Neurospora* where 75 temporally regulated phosphorylation sites have been uncovered where phosphorylation either stabilized or destabilized FRQ (Baker et al. 2009). Therefore, it is possible that CK2-dependent phosphorylation of CCA1 may lead to activation or repression of activity based on the temporal and spatial conditions, as well as the sites of phosphorylation. Additional studies are required to further characterize the phase-specific phosphorylation of CCA1 and the impacts of changes in temperature.

It appears that the interplay among multiple regulatory loops is important for proper buffering in response to temperature changes, and this involves a cascade of molecular players as evident from both genetic and natural variation studies. Ultimately, the controlled regulation of period lengtheners and shorteners, for example, CCA1/LHY and PRR7/PRR9, respectively, is necessary for proper temperature compensation. The type of regulation imposed on these factors appears to be dependent on the direction of the temperature change (Figure 6.2C).

6.3 Temperature entrainment

Following the discovery of an endogenous circadian clock, inquiries arose about the properties of the clock and whether inputs from the environment are in fact necessary for maintaining the correct circadian oscillations. As early as 1832, de Candolle (de Candolle 1832) established that the rhythm in leaf movement could be inverted by reversing the alternation of light and dark, establishing that the clock could be entrained to new environmental cycles. Subsequent studies emphasized the role of light as an obvious *zeitgeber*. Light pulses were capable of shifting the phase of the clock. Moreover, the magnitude and direction (advances *versus* delays) of the resultant phase shift was dependent on the time of day at which the pulse was given, indicating that the clock modulates its own sensitivity to phase-shifting stimuli (Pittendrigh et al. 1958; Pittendrigh and Bruce 1959). In this way, the term 'temperature entrainment' addresses how temperature acts as a *zeitgeber* to set the phase of the clock. The ability to entrain the circadian clock by temperature also offers a way to delineate the contributions of clock components by examining their response to light *versus* temperature cycles.

Surprisingly, it appears that thermal entrainment is under distinct genetic regulation compared to photic entrainment. A comparison of photic and thermal entrainment on an *Arabidopsis* RIL population revealed nonoverlapping QTL for the two entrainment conditions, suggesting distinct allelic control (Boikoglou et al. 2011). In order to demonstrate temperature entrainment, organisms are grown under LL with temperature cycles. In the case of *Arabidopsis*, an example of a thermocycle commonly used is 12 h at 22°C and 12 h at 16°C (Boikoglou et al. 2011). Although a temperature difference of as little as 4°C has been shown to be sufficient for temperature entrainment (Michael et al. 2003a), the lower limit of this sensitivity has not been determined. Plants entrained in thermocycles and shifted to constant temperature continued to oscillate based on the original temperature phase as observed with light entrainment (Somers et al. 1998). The phase of entrainment in response to temperature cycles is typically the warm temperature that is interpreted as the light part of the cycle, such that light-expressed genes, including *CCA1* and *LHY*, peak early in the warm part of the cycle. Period mutants that properly entrain to temperature cycles show the expected phase relationships where short-period mutants exhibited a leading phase and long-period mutants showed a lagging phase during entrainment (Salomé and McClung 2005). Mutants that were deficient in temperature entrainment did not anticipate the warm–cold or cold–warm transition. The *toc1-2* and *lhy-20* short-period mutants showed a leading phase after 5 days at 22°C LD entrainment with release into 22°–18°C thermocycles in phase with the preceding photocycles (22°C replaces the light; 18°C replaces the dark). Similarly, the *ztl-4* long-period mutant exhibited a lagging phase upon transfer to thermocycles (Salomé and McClung 2005). These results suggested that these mutants were affected in their ability to be entrained by both light and temperature cycles.

Further examination of other clock mutants revealed a role for members of the PRR family in temperature entrainment. The *prr7-3*, *prr5-3*, and to a lesser extent *prr9-1* single mutants all showed defects in their ability to entrain to thermocycles. The *prr3-1* mutant was able to entrain to thermocycles after being initially entrained to photocycles (Salomé and McClung 2005). These entrainment assays were performed using leaf movement data, and given the discrepancies that have been observed in the *GI* alleles (Gould et al. 2006; Salomé et al. 2010) comparing leaf movement and luciferase activity, further studies using luciferase reporters in these *PRR* mutants are needed.

Strikingly, the *prr7-3 prr9-1* double mutant was found to have an extremely long period of approximately 30–36 h (Farré et al. 2005; Salomé and McClung 2005). The double mutant following temperature entrainment of hot/cold, LL into hot/hot, and LL (HC, LL into HH, LL) exhibited arhythmicity in cotyledon movement and *CCA1* or *TOC1* promoter luciferase experiments. Occasionally the first peak of *CCA1* or *TOC1* expression occurred but was not maintained (Salomé and McClung 2005; Salomé et al. 2010). Since the *prr7-3 prr9-1* mutant can be entrained by LD cycles (Farré et al. 2005; Salomé and McClung 2005), the response to temperature could be tested by transferring photocycle-entrained plants to a HC and LL cycle that is in phase with the photocycle. Under these entrainment conditions, *CCA1* and *LHY* promoter expression peaked later than wild type but showed mild anticipation of the temperature transition. The *TOC1* promoter showed a different expression pattern with a peak 12 h earlier than in wild type, and no anticipation of the temperature transition was observed (Salomé and McClung 2005). Not surprisingly, the triple *cca1 lhy toc1* and quadruple *prr5 prr7 prr9 toc1* mutants were unable to entrain to LD or HC cycles (Ding et al. 2007; Yamashino et al. 2008). This emphasized the importance of PRR7 and PRR9 regulation of *CCA1* and *LHY* expression as well as a connection between proper phasing of *TOC1* expression and the ability to maintain temperature entrainment.

Two additional members of the PRR family that have not been thoroughly characterized, but may be important in temperature integration into the clock, are PRR3 and PRR5. Salomé and McClung (2005) included the single mutants of these genes in their temperature entrainment assays, and it was clear that *prr3-1* showed a mild lagging phase phenotype following thermocycle entrainment. However, *prr5-3* had a strong entrainment phenotype, especially when seedlings entrained by photocycles were transferred into a thermocycle (HH, LD into HC, LL), suggesting a defect in the ability to be entrained by temperature. Given the similar trend in phenotypes in these two mutants, it would be worthwhile to test temperature entrainment in the *prr3 prr5* double mutant. The *PRR* genes each showed peak expression at different times of day. *PRR9* peaked at dawn, *PRR7* in the morning, *PRR5* at midday, and *PRR3* and *TOC1* at the end of the day (Matsushika et al. 2000; Mizuno 2004). Although *PRR5*, *PRR7*, and *PRR9* are thought to have some redundant function, their single-mutant phenotypes, as well as their expression patterns, suggested that they contribute differentially to the temperature-response mechanism

(Mizuno 2004; Nakamichi et al. 2010). To define the contribution of these proteins will require further studies, in particular information on the abundance and activities of these proteins during temperature entrainment.

Further insight into temperature entrainment has arisen through the studies of ELF3, a member of the EC. Loss of ELF3 function conferred a number of phenotypes, including early flowering, photoperiod insensitivity, altered photomorphogenesis, and a long hypocotyl under white light (Zagotta et al. 1992). All of these phenotypes involve processes that are tightly linked to the circadian clock (Yakir et al. 2007). Leaf movement analysis in *elf3* did not detect rhythmic movement under LL. A luciferase reporter using the promoter of chlorophyll *a*/*b* binding protein gene (*CAB2*) confirmed the lack of a rhythm in *elf3* under red and blue light (Hicks et al. 1996b). The *elf3* mutant was disrupted in temperature entrainment and was unable to maintain rhythmicity following 4°C or 10°C thermocycle differences (McWatters et al. 2000; Thines and Harmon 2010).

Wild-type plants that are entrained to 24 h thermocycles and transferred to shorter cycles, for example, 6 h warm/6 h cold, show a 24 h periodicity where every other peak is greater than the 6 h peak. This type of experiment is known as frequency demultiplication and tests whether the plant is simply responding to the alternation between warm and cold (LD) or instead is still manifesting a functioning endogenous circadian oscillator. Mutants with nonfunctional oscillators do not show any signs of their previous 24 h entrainment and instead peak in phase with alternating external cues (McWatters et al. 2000; McWatters et al. 2007). This was observed for *elf3* under 12 h thermocycles, demonstrating the importance of ELF3 for proper temperature entrainment (Thines and Harmon 2010). Interestingly, overexpression of ELF3 did not affect the phase response following resetting by temperature cues, implying that ELF3 was not important for the ability of the clock to perceive temperature (Thines and Harmon 2010). This observation not only established ELF3 as a target of temperature input to the clock but also distinguishes mechanisms of perception and response to temperature.

The temperature induction of *PRR7*, *PRR9*, and *GI* mRNA abundance was eliminated in the *elf3* mutant, and the basal levels of expression of these genes were elevated, especially at night (Thines and Harmon 2010). The elevated expression at night was consistent with the normal role of the EC in gating the induction of these genes by repressing expression at night (Nusinow et al. 2011). It appears that the role of ELF3 is not specific to temperature entrainment, but, rather, ELF3 is necessary to maintain proper circadian clock function (Herrero et al. 2012) and regulation of temperature-responsive clock genes (Thines and Harmon 2010).

6.4 Cold tolerance

Temperate plants rely on seasonal cues from the environment, such as day length and temperature, to signal impending changes in growth conditions (Rohde and Bhalerao 2007; de Montaigu et al. 2010). Cold plays a major role throughout the plant life cycle from stratification, the cold-induced breaking of seed dormancy, to vernalization, the cold-induced process that conditions the plant to respond to flower-inducing stimuli. In order to survive the winter season, plants must undergo a process called cold acclimation. The gradual decline in temperature exposes the plant to nonfreezing temperatures. This triggers an increase in cold tolerance, allowing the plant to survive temperatures that would be lethal without this preconditioning. Due to the pressures to maximize the growing season for crops despite the threat of freezing temperatures, considerable attention is focused on understanding how plants respond to the cold (Thomashow 1999).

As highlighted in Chapters 2 and 5, the cold-response pathway involves three members of the AP2/ERF family of transcription factors referred to as CBF1–CBF3/DREB1B, DREB1C, and DREB1A. These genes are tandemly linked in *Arabidopsis* and induced within 15 min of exposure to low temperature (Stockinger et al. 1997; Gilmour et al. 1998; Liu et al. 1998). CBFs bind to C-repeat/dehydration-responsive elements (CRT/DRE) in the promoters of more than 100 cold-responsive genes, commonly referred to as the CBF regulon (Shinozaki and Yamaguchi-Shinozaki 2000; Vogel et al. 2005). The induction of these CBF-targeted genes is sufficient to induce freezing tolerance, because constitutive expression of the CBF genes induced freezing tolerance in the absence of a low-temperature treatment (Jaglo-Ottosen et al. 1998; Liu et al. 1998). Among the cold-induced genes are the *COR* genes that encode hydrophilic polypeptides, in particular *COR15a*, that are important for the cryoprotection of the plasma membrane during freeze tolerance (Steponkus et al. 1998). The CBF pathway is also involved in cold-induced changes in the metabolome that lead to the synthesis of cryoprotectants (Cook et al. 2004; Kaplan et al. 2004).

Microarray studies of the circadian-regulated transcriptome in *Arabidopsis* revealed a circadian oscillation in the pattern of expression of *CBF3* with a peak at midday (Harmer et al. 2000). Later studies showed that *CBF1* and *CBF2* are also circadian regulated, along with two other cold-responsive genes, *RAV1* and *ZAT12*, suggesting a role for the clock in modulating the response to low temperature (Fowler et al. 2005). The consequence of the circadian regulation of the *CBF* genes is a time-of-day-dependent effect on low-temperature-induced transcript accumulation, shown to result in an increase in cold induction during the day compared to the night (Fowler et al. 2005; Dong et al. 2011). The rhythmicity of *CBF1* and *CBF3*, as well as several *COR* genes, was lost in the *cca1 lhy* double mutant, while the

CBF2 amplitude was dramatically reduced. The effect of these disruptions in cold-responsive gene expression was a 50% reduction in freezing tolerance in *cca1 lhy* based on electrolyte leakage and seedling survival assays (Espinoza et al. 2010; Dong et al. 2011).

Consistent with the role of the PRRs in the temperature-compensation mechanism, the *prr5 prr7 prr9* mutant was shown to be significantly more resistant to freezing tolerance than wild type such that 62% of the mutants survived a 1-day −5°C cold treatment that was completely lethal to wild-type plants (Nakamichi et al. 2009). Not surprisingly, this arrhythmic mutant was affected in many processes regulated by the clock such as resistance to salt and drought stress as well as misexpression of a large suite of genes involved in a range of processes from abiotic and biotic stress to metabolism and development (Nakamichi et al. 2009).

Evidently, the circadian clock tightly controls the cold-response pathway as seen by the phenotypes of these clock mutants. However, another important question to address is how the clock itself responds to low temperature. Bieniawska et al. (2008) showed that in *Arabidopsis*, oscillations in steady-state mRNA abundance for many clock and clock-regulated genes were dramatically compromised at 4°C. For most transcripts, the rhythmic amplitude was greatly attenuated in LD and virtually abolished in LL. Abundance of most, but not all, transcripts dampened to high levels. The transcript of the clock gene *LUX* retained a robust oscillation at 4°C in LD but showed only low amplitude oscillations in LL, and because the experiment only examined the first two cycles in LL, it was unclear if that apparent rhythmicity persisted. Not surprisingly, the cold-responsive genes that have been shown to be clock regulated, such as *CBF1* and *CBF2*, are induced following cold treatment and remain elevated, becoming arrhythmic in both LD and LL (Bieniawska et al. 2008). Thus, although the clock plays a crucial role gating the transcriptional response to cold, clock function is greatly compromised if not completely abolished at 4°C, with the acknowledged caveat that

Bieniawska et al. (2008) only examined the first two cycles following onset of cold. The ability of *Arabidopsis* to persist for extended periods at cold temperatures raises the question of whether clock function might recover to some extent during extended cold treatment.

In addition to being involved in the response to a rapid drop in temperature, the circadian clock is also important under extended cold conditions to enable proper endodormancy in trees such as chestnut, *Castanea sativa* Mill. (Ramos et al. 2005). Endodormancy is the deep dormancy during winter that inhibits growth until a chilling requirement is reached. Short-day conditions and low temperatures lead to growth cessation, bud set, and the start of cold acclimation. Stem tissue collected from chestnut at midday in June and December showed *TOC1* transcript in the winter sample at a time when *TOC1* expression is normally inhibited. Further analysis revealed oscillations of *CCA1* and *LHY* mRNA under constant conditions, confirming that a circadian rhythm existed at 20°C; however, similar to *Arabidopsis*, the oscillations were abolished in the winter. This effect was not due to the endodormancy state since cold pulses on nondormant chestnut seedlings produced the same effect (Ramos et al. 2005). *PRR* homologues were identified in chestnut based on sequence similarity and expression pattern and designated *CsPRR5*, *CsPRR7*, and *CsPRR9*. Consistent with the other clock genes, the *CsPRRs* were highly expressed in winter samples with no oscillation pattern evident. Rhythmicity was regained following 1 week of warm temperatures, demonstrating a clear connection between temperature sensing, circadian clock response, and regulated plant growth (Ibanez et al. 2008).

Analysis of the metabolome suggested that a subset of metabolites showed circadian regulation based on their accumulation in LL; these included maltose, O-acetyl serine, valine, and alanine. Of these, only maltose maintained oscillation patterns at 4°C (Espinoza et al. 2010). Network modeling based on correlations between changes in metabolites under diurnal (LD) and circadian (LL) conditions at 4°C and 20°C revealed more connections at 4°C in LL and LD, implying a large coordinated shift in the metabolic profile at 4°C (Espinoza et al. 2010). In addition to changes in metabolite levels, cytosolic Ca^{2+} ($[Ca^{2+}]_{cyt}$) levels also responded to changes in temperature, and this was also time-of-day dependent. A low-temperature treatment of 0°C resulted in an increase in guard cell $[Ca^{2+}]_{cyt}$ that was greatest at midday (ZT6.5 h) (Dodd et al. 2006). This increase may be important for the regulation of the CBF response through a calcium-dependent calmodulin activation of CAMTA3, a regulator of *CBF1*, *CBF3*, and *ZAT12*. CAMTA3, a member of the calmodulin-binding transcription activator family, has been shown to bind to the promoters of *CBF1*, *CBF3*, and *ZAT12* and positively regulate their expression in response to cold treatment. The *camta1 camta3* double mutant showed a reduction in freezing tolerance following cold acclimation, and *CBF2* transcript levels were reduced by 50% in the *camta3* mutants following cold treatment, suggesting that this is one of several regulators of *CBF2* (Doherty et al. 2009). This provides a possible mechanism for the integration of calcium signaling through CAMTA3 and an important component to the process of freezing tolerance.

Freezing tolerance is particularly relevant to the growth and optimization of tomato, an extremely chilling-sensitive crop. Upon transfer to low temperatures, tomato plants showed a sharp reduction in chloroplast function, leading to a decrease in photosynthesis due to limits on carbon availability. Cold treatment also resulted in a suspension in circadian regulation of transcription (Martino-Catt and Ort 1992). Interestingly, the

CBF pathway is slightly modified in *Solanum* species. Although the *CBF* genes show a similar structural organization, the locus is fairly divergent across species. Tomato (*Solanum lycopersicum*) has *CBF1–3* tandemly arranged similarly to *Arabidopsis*; however, only *Sl-CBF1* was cold responsive and this is consistent with the high conservation of the upstream region of *CBF1* in the *Solanum* family. *Solanum CBF1* also displays circadian regulation (Pennycooke et al. 2008). Other *Solanum* species such as *Solanum tuberosum* (potato) and *Solanum commersonii* (diploid tuber-bearing potato) have copies of *CBF4* that are cold responsive. The more freezing-tolerant *S. commersonii* showed higher expression levels of *CBF1* and *CBF4* following cold temperatures compared to the more sensitive *S. tuberosum*, demonstrating a clear connection between the regulation of the *CBFs* and the plant's ability to tolerate low temperatures (Pennycooke et al. 2008).

Aside from freezing tolerance, many plants rely on the cold during the winter season to know when to flower and require vernalization to stimulate flowering in the spring (discussed in detail in Chapter 4). To prevent unwanted flowering following a brief temperature drop in the fall, vernalization requires an extended cold treatment before initializing flowering. In *Arabidopsis*, two important genes required for proper flowering time control are *FLC* and *FRIGIDA (FRI)*. *FLC* is an inhibitor of flowering and its expression is promoted by *FRI* in the fall and inhibited by vernalization (Amasino 2010). Due to the effect of altered *FLC* expression and period length, Salathia et al. (2006) examined the effect of vernalization on the circadian clock. Leaf movement analysis in various mutant backgrounds demonstrated that vernalization shortened circadian period independent of *FLC* and *FRI*, suggesting a connection between vernalization and the circadian clock (Salathia et al. 2006). One well-established result of vernalization is epigenetic changes in chromatin structure at the *FLC* locus (Kim et al. 2009; Amasino 2010; Crevillén and Dean 2011). Accumulating data indicate that rhythmic chromatin changes are associated with cycling clock gene expression (Perales and Más 2007; Jones et al. 2010; Farinas and Mas 2011; Lu et al. 2011). One possibility is that vernalization induces analogous epigenetic changes at one or more clock gene loci, with consequent changes in clock period (Sheldon et al. 2009; Deng et al. 2011). These changes in clock period under constant conditions translate to a phase shift under diurnal conditions. The clock regulates the phase of the response relative to the LD cycle, resulting in gated control of output pathways as demonstrated by the distinct phases of peak hypocotyl growth in LL and LD conditions (Dowson-Day and Millar 1999; Nozue et al. 2007). Although period estimates are important for drawing conclusions about the involvement of the circadian clock, the associated phase is directly applicable to the diurnal growth of the plant seen in nature.

6.5 Splicing

The role of AS in the circadian clock is becoming increasingly recognized and evidence is accumulating for thermosensitivity of splicing events. In *Neurospora*, two forms of the clock protein FRQ generated via AS are present at ratios that are temperature dependent. These AS events are thought to be important for controlling the response to ambient temperature (Liu et al. 1997; Colot et al. 2005; Diernfellner et al. 2007). Similarly, the *PER* gene in *Drosophila* is alternatively spliced at low temperature, leading to an increase in PER protein earlier in the day than at normal temperatures (Majercak et al. 1999).

Recently, thermosensitive AS has been demonstrated in *Arabidopsis* for a number of clock genes. Two alternatively spliced isoforms with premature termination codons for *CCA1* accumulated differentially at low temperature (Filichkin et al. 2010; Filichkin and Mockler 2012). James et al. (2012) extended this analysis using a high-resolution RT-PCR system to detect AS events for *CCA1*, *LHY*, *TOC1*, *PRR3*, *PRR5*, *PRR7*, *PRR9*, *GI*, *ZTL*, and *CHE*. They identified 63 new splicing events for these clock genes, several of which were shown to be temperature dependent. In the case of *CCA1* and *LHY*, a shift from 20°C to 4°C at dusk resulted in an increase of unproductive AS products for *LHY* and a decrease for *CCA1*. Consistent with these results, CCA1 protein levels were much higher than those of LHY following a shift to cold temperatures relative to 20°C. Similar opposing patterns of AS product accumulation were observed for *PRR7* and *PRR9* (James et al. 2012). Although one splicing factor, PRMT5, has been shown to be important for the AS of *PRR7* and *PRR9* (Deng et al. 2010; Hong et al. 2010; Sanchez et al. 2010), it is not known whether this process is temperature sensitive. The exact consequence of these AS events on protein abundance has not been thoroughly explored; future studies examining protein accumulation at various temperatures are needed.

Seo et al. (2012) proposed a mechanism for CCA1 protein regulation through an alternatively spliced *CCA1* transcript containing an early stop codon in the retained fourth intron. They suggested that an ATG following the premature stop codon leads to translation of a truncated protein, CCA1ß, that lacks the MYB DNA-binding domain but retains the dimerization domain necessary for interaction with CCA1 and LHY. This truncated CCA1ß protein interacted with full-length CCA1 and LHY in yeast and inhibited target promoter binding. Overexpression of CCA1ß *in planta* shortened circadian period, consistent with the truncated CCA1ß acting as a negative regulator of CCA1 and LHY activity; however, it has yet to be demonstrated directly that this truncated CCA1ß protein is present in plants (Seo et al. 2012). Consistent with previous studies (Gould et al. 2006), Seo et al. (2012) confirmed the temperature-dependent ratio of

CCA1 splice transcripts with a decrease in the inhibitory splice variant at low temperature, suggesting greater activity of CCA1 protein.

The presence of nonfunctional transcripts is important to consider when interpreting measurements of total transcript levels. In the case of James et al. (James et al. 2012), circadian oscillations were maintained at 4°C and transcript levels for *CCA1* and *LHY* were reduced, whereas Bieniawska et al. (2008) showed that the oscillations were severely reduced but transcript levels remain high. Other discrepancies in the pattern of expression of clock genes in response to temperature changes have been mentioned previously, highlighting the complexity of the transcriptional regulation of these genes. Future studies on the functional implications of temperature on circadian clock genes need to address both transcript and protein level.

6.6 Concluding remarks

The impending threat of climate change has become a driving force for research into how plants sense and respond to temperature. The circadian clock is an important regulator of plant growth and likely an essential driver of the temperature response due to the role of the clock in the coordination of important metabolic, physiological, and behavioral processes with the external environment. It has been recently shown that metabolism and growth are temperature compensated, although the mechanistic details remain incompletely defined. Biomass accumulation, starch turnover, and protein content were all unaltered with decreases in the night temperature. Variation in the overall growth rate was dependent on changes in the daytime temperature (Pyl et al. 2012). The inherent property of temperature compensation and the ability of the clock to be entrained to temperature cycles highlight the close relationship that exists between changes in ambient temperature and clock adaptability.

An important question remains: how does the clock sense a change in temperature? There are several mechanisms that contribute to the thermal response, some of which are depicted in Figure 6.3. For example, increases in temperature lead to a reduction in H2A.Z nucleosome occupancy causing a relaxation in chromatin, allowing for the expression of warm temperature-responsive genes, such as *HSP70* (Kumar and Wigge 2010). This heat-sensitive response may trigger the activation of circadian genes; the mechanism that exists for *TOC1* activation by H3 acetylation might be temperature sensitive. It seems likely that the activation (or inactivation) of temperature-sensitive enzymes like CK2 will be important for triggering a fast response through posttranslational activation (or inactivation) of clock proteins as seen with CCA1. Finally, the role of AS is an important avenue for further investigation. Low-temperature effects on *LHY* are apparent,

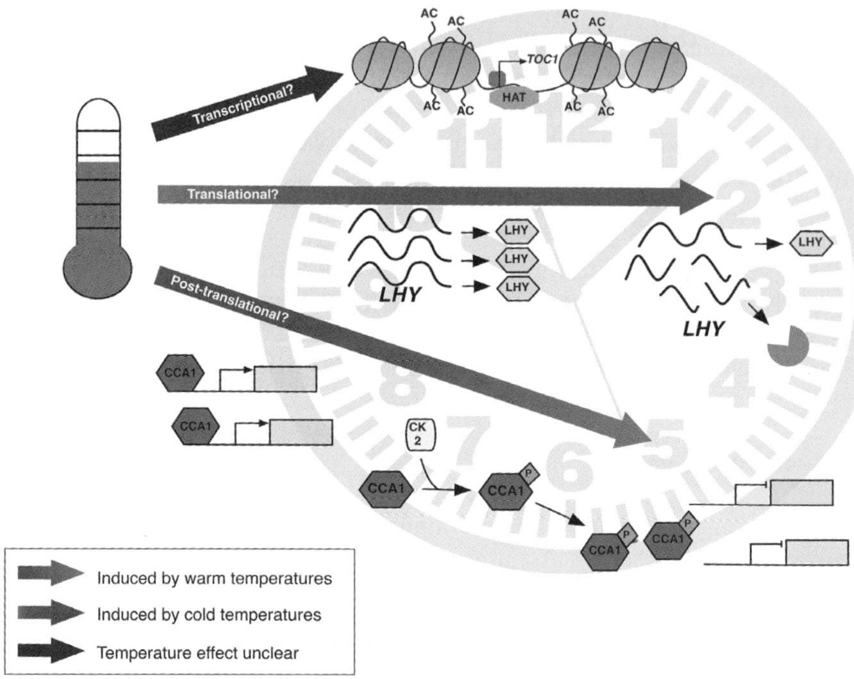

Figure 6.3 Possible points of temperature integration into the circadian clock. It is still not clear how the clock senses changes in ambient temperature. The three processes depicted in this figure integrate temperature at the level of transcriptional, translational, and posttranslational regulation. The *black arrow* represents the transcriptional regulation of chromatin remodeling that has been shown to be important for *TOC1* regulation but not yet linked to temperature. However, high temperature has been shown to deplete H2A.Z nucleosome occupancy leading to activation of the warm temperature transcriptome (Kumar and Wigge 2010). The translational arrow describes the temperature-dependent AS events for *LHY*. As the temperature cools (indicated by a transition from *red* to *blue* in the *arrow*), there is an increase in unproductive splice forms for *LHY* that are likely responsible for the decrease in protein abundance. Finally, the posttranslational arrow depicts the high-temperature-dependent (illustrated by a *blue* to *red* color transition) phosphorylation of CCA1 by CK2, leading to a decrease in CCA1 activity. The exact timing or sensitivity of this response is unclear but may be an important mechanism for fine-tuning the temperature response. For color detail, please see color plate section.

and it will be important to examine other clock genes at various temperatures to assess the implications of this regulation on protein abundance. Figure 6.3 does not address $[Ca^{2+}]_{cyt}$ although it has been known for many years that changes in $[Ca^{2+}]_{cyt}$ are among the most rapid events detected following exposure to either low or high temperatures and are likely to be close to at least one primary temperature-sensing mechanism (Knight et al. 1996; Gong et al. 1998; McClung and Davis 2010). Nonetheless, within the context of the circadian clock, it seems that $[Ca^{2+}]_{cyt}$ is used as a clock output signaling intermediate rather than as a component of an input pathway signaling from a primary temperature sensor (Dodd et al. 2005a; Xu et al. 2007).

Significant gaps remain in our understanding of plant temperature responses and, in particular, how a clock can be temperature responsive and entrained by temperature cycles yet also buffered (compensated) against changes in temperature. Neither the primary temperature sensors nor the signaling pathways to the clock have been fully elucidated. Future studies must face the challenges of both identifying temperature signaling components and deciphering the order of events following a change in temperature in terms of the spatial and temporal regulation of the clock network.

Acknowledgments

This work was supported by grants (IOS-0923752 and IOS-1029565) from the National Science Foundation. Kathleen Greenham was supported by the National Science Foundation Postdoctoral Research Fellowship in Biology under grant No. IOS-1202779. We would like to thank Colleen Doherty and Seth Davis for critical reading of the manuscript.

References

Alabadí D, Oyama T, Yanovsky MJ, Harmon FG, Más P, Kay SA (2001) Reciprocal regulation between *TOC1* and *LHY/CCA1* within the *Arabidopsis* circadian clock. Science 293: 880–883.

Amasino R (2010) Seasonal and developmental timing of flowering. Plant J. 61: 1001–1013.

Baker CL, Kettenbach AN, Loros JJ, Gerber SA, Dunlap JC (2009) Quantitative proteomics reveals a dynamic interactome and phase-specific phosphorylation in the Neurospora circadian clock. Mol. Cell 34: 354–363.

Barak S, Tobin EM, Andronis C, Sugano S, Green RM (2000) All in good time: The *Arabidopsis* circadian clock. Trends Plant Sci. 5: 517–522.

Bell-Pedersen D, Cassone VM, Earnest DJ, Golden SS, Hardin PE, Thomas TL, Zoran MJ (2005) Circadian rhythms from multiple oscillators: Lessons from diverse organisms. Nat. Rev. Genet. 6: 544–556.

Bieniawska Z, Espinoza C, Schlereth A, Sulpice R, Hincha DK, Hannah MA (2008) Disruption of the *Arabidopsis* circadian clock is responsible for extensive variation in the cold-responsive transcriptome. Plant Physiol. 147: 263–279.

Bodenstein C, Heiland I, Schuster S (2011) Calculating activation energies for temperature compensation in circadian rhythms. Phys. Biol. 8: 056007.

Boikoglou E, Ma Z, von Korff M, Davis AM, Nagy F, Davis SJ (2011) Environmental memory from a circadian oscillator: The *Arabidopsis thaliana* clock differentially integrates perception of photic versus thermal entrainment. Genetics 189: 655–664.

Bruce VG, Pittendrigh CS (1956) Temperature independence in a unicellular "clock". Proc. Natl. Acad. Sci. USA 42: 676–682.

de Candolle AP (1832) Physiologie Végétale. Bechet Jeune, Paris.

Chow BY, Helfer A, Nusinow DA, Kay SA (2012) ELF3 recruitment to the *PRR9* promoter requires other Evening Complex members in the *Arabidopsis* circadian clock. Plant Signal. Behav. 7: 1–4.

Colot HV, Loros JJ, Dunlap JC (2005) Temperature-modulated alternative splicing and promoter use in the circadian clock gene *frequency*. Mol. Biol. Cell 16: 5563–5571.

Cook D, Fowler S, Fiehn O, Thomashow MF (2004) A prominent role for the CBF cold response pathway in configuring the low-temperature metabolome of *Arabidopsis*. Proc. Natl. Acad. Sci. USA 101: 15243–15248.

Crevillén P, Dean C (2011) Regulation of the floral repressor gene *FLC*: The complexity of transcription in a chromatin context. Curr. Opin. Plant Biol. 14: 38–44.

Daniel X, Sugano S, Tobin EM (2004) CK2 phosphorylation of CCA1 is necessary for its circadian oscillator function in *Arabidopsis*. Proc. Natl. Acad. Sci. USA 101: 3292–3297.

Deng X, Gu L, Liu C, Lu T, Lu F, Lu Z, Cui P, Pei Y, Wang B, Hu S, Cao X (2010) Arginine methylation mediated by the *Arabidopsis* homolog of PRMT5 is essential for proper pre-mRNA splicing. Proc. Natl. Acad. Sci. USA 107: 19114–19119.

Deng W, Ying H, Helliwell CA, Taylor JM, Peacock WJ, Dennis ES (2011) FLOWERING LOCUS C (FLC) regulates development pathways throughout the life cycle of *Arabidopsis*. Proc. Natl. Acad. Sci. USA 108: 6680–6685.

Diernfellner A, Colot HV, Dintsis O, Loros JJ, Dunlap JC, Brunner M (2007) Long and short isoforms of Neurospora clock protein FRQ support temperature-compensated circadian rhythms. FEBS Lett. 581: 5759–5764.

Ding Z, Doyle MR, Amasino RM, Davis SJ (2007) A complex genetic interaction between *Arabidopsis thaliana* TOC1 and CCA1/LHY in driving the circadian clock and in output regulation. Genetics 176: 1501–1510.

Dixon LE, Knox K, Kozma-Bognar L, Southern MM, Pokhilko A, Millar AJ (2011) Temporal repression of core circadian genes Is mediated through EARLY FLOWERING 3 in *Arabidopsis*. Curr. Biol. 21: 120–125.

Dodd AN, Love J, Webb AAR (2005a) The plant clock shows its metal: Circadian regulation of cytosolic free Ca^{2+}. Trends Plant Sci. 10: 15–21.

Dodd AN, Salathia N, Hall A, Kevei E, Toth R, Nagy F, Hibberd JM, Millar AJ, Webb AAR (2005b) Plant circadian clocks increase photosynthesis, growth, survival, and competitive advantage. Science 309: 630–633.

Dodd AN, Jakobsen MK, Baker AJ, Telzerow A, Hou S-W, Laplaze L, Barrot L, Poethig RS, Haseloff J, Webb AAR (2006) Time of day modulates low-temperature Ca^{2+} signals in *Arabidopsis*. Plant J. 48: 962–973.

Doherty CJ, Kay SA (2010) Circadian control of global gene expression patterns. Annu. Rev. Genet. 44: 419–444.

Doherty CJ, Van Buskirk HA, Myers SJ, Thomashow MF (2009) Roles for *Arabidopsis* CAMTA transcription factors in cold-regulated gene expression and freezing tolerance. Plant Cell 21: 972–984.

Dong MA, Farré EM, Thomashow MF (2011) CIRCADIAN CLOCK-ASSOCIATED 1 and LATE ELONGATED HYPOCOTYL regulate expression of the C-REPEAT BINDING FACTOR (CBF) pathway in *Arabidopsis*. Proc. Natl. Acad. Sci. USA 108: 7241–7246.

Dowson-Day MJ, Millar AJ (1999) Circadian dysfunction causes aberrant hypocotyl elongation patterns in *Arabidopsis*. Plant J. 17: 63–71.

Doyle MR, Davis SJ, Bastow RM, McWatters HG, Kozma-Bognar L, Nagy F, Millar AJ, Amasino RM (2002) The *ELF4* gene controls circadian rhythms and flowering time in *Arabidopsis thaliana*. Nature 419: 74–77.

Easterling WE, Aggarwal PK, Batima P, Brander KM, Erda L, Howden SM, Kirilenko A, Morton J, Soussana J-F, Schmidhuber J, Tubiello FN (2007) Food, fibre and forest products. In: Parry ML, Canziani OF, Palutikof JP, van der Linden PJ, Hanson CE (eds). Climate Change 2007: Impacts, adaptation and vulnerability. Contribution of Working Group II to the Fourth Assessment Report of the Intergovernmental Panel on Climate Change. Cambridge University Press, Cambridge, pp. 273–313.

Edgar RS, Green EW, Zhao Y, van Ooijen G, Olmedo M, Qin X, Xu Y, Pan M, Valekunja UK, Feeney KA, Maywood ES, Hastings MH, Baliga NS, Merrow M, Millar AJ, Johnson CH, Kyriacou CP, O'Neill JS, Reddy AB (2012) Peroxiredoxins are conserved markers of circadian rhythms. Nature 485: 459–464.

Edwards KD, Lynn JR, Gyula P, Nagy F, Millar AJ (2005) Natural allelic variation in the temperature compensation mechanisms of the *Arabidopsis thaliana* circadian clock. Genetics 170: 387–400.

Edwards KD, Anderson PE, Hall A, Salathia NS, Locke JCW, Lynn JR, Straume M, Smith JQ, Millar AJ (2006) *FLOWERING LOCUS C* mediates natural variation in the high-temperature response of the *Arabidopsis* circadian clock. Plant Cell 18: 639–650.

Espinoza C, Degenkolbe T, Caldana C, Zuther E, Leisse A, Willmitzer L, Hincha DK, Hannah MA (2010) Interaction with diurnal and circadian regulation results in dynamic metabolic and transcriptional changes during cold acclimation in *Arabidopsis*. PLoS One 5: e14101.

Farinas B, Mas P (2011) Functional implication of the MYB transcription factor RVE8/LCL5 in the circadian control of histone acetylation. Plant J. 66: 318–329.

Farré EM, Harmer SL, Harmon FG, Yanovsky MJ, Kay SA (2005) Overlapping and distinct roles of PRR7 and PRR9 in the *Arabidopsis* circadian clock. Curr. Biol. 15: 47–54.

Filichkin SA, Mockler TC (2012) Unproductive alternative splicing and nonsense mRNAs: A widespread phenomenon among plant circadian clock genes. Biol. Direct 7: 20.

Filichkin SA, Priest HD, Givan SA, Shen R, Bryant DW, Fox SE, Wong W-K, Mockler TC (2010) Genome-wide mapping of alternative splicing in *Arabidopsis thaliana*. Genome Res. 20: 45–58.

Foster RG, Roenneberg T (2008) Human responses to the geophysical daily, annual and lunar cycles. Curr. Biol. 18: R784–R794.

Fowler S, Lee K, Onouchi H, Samach A, Richardson K, Morris B, Coupland G, Putterill J (1999) *GIGANTEA*: A circadian clock-controlled gene that regulates photoperiodic flowering in *Arabidopsis* and encodes a protein with several membrane-spanning domains. EMBO J. 18: 4679–4688.

Fowler SG, Cook D, Thomashow MF (2005) Low temperature induction of *Arabidopsis* CBF1, 2, and 3 is gated by the circadian clock. Plant Physiol. 137: 961–968.

Gardner GF, Feldman JF (1981) Temperature compensation of circadian period length in clock mutants of *Neurospora crassa*. Plant Physiol. 68: 1244–1248.

Gendron JM, Pruneda-Paz JL, Doherty CJ, Gross AM, Kang SE, Kay SA (2012) *Arabidopsis* circadian clock protein, TOC1, is a DNA-binding transcription factor. Proc. Natl. Acad. Sci. USA 109: 3167–3172.

Gilmour SJ, Zarka DG, Stockinger EJ, Salazar MP, Houghton JM, Thomashow MF (1998) Low temperature regulation of the *Arabidopsis* CBF family of AP2 transcriptional activators as an early step in cold-induced COR gene expression. Plant J. 16: 433–442.

Gong M, van der Luit AH, Knight MR, Trewavas AJ (1998) Heat-shock-induced changes in intracellular Ca^{2+} level in tobacco seedlings in relation to thermotolerance. Plant Physiol. 116: 429–437.

Goodspeed D, Chehab EW, Min-Venditti A, Braam J, Covington MF (2012) *Arabidopsis* synchronizes jasmonate-mediated defense with insect circadian behavior. Proc. Natl. Acad. Sci. USA 109: 4674–4677.

Gould PD, Locke JCW, Larue C, Southern MM, Davis SJ, Hanano S, Moyle R, Milich R, Putterill J, Millar AJ, Hall A (2006) The molecular basis of temperature compensation in the *Arabidopsis* circadian clock. Plant Cell 18: 1177–1187.

Greenbaum D, Colangelo C, Williams K, Gerstein M (2003) Comparing protein abundance and mRNA expression levels on a genomic scale. Genome Biol. 4: 117.

Halberg F, Halberg E, Barnum CP, Bittner JJ (1959) Physiologic 24-hour periodicity in human beings and mice, the lighting regimen and daily routine. In: Withrow RB (ed). Photoperiodism and related phenomena in plants and animals. American Association for the Advancement of Science, Washington, DC, pp. 803–878.

Han L, Mason M, Risseeuw EP, Crosby WL, Somers DE (2004) Formation of an SCF^{ZTL} complex is required for proper regulation of circadian timing. Plant J. 40: 291–301.

Harmer SL (2009) The circadian system in higher plants. Annu. Rev. Plant Biol. 60: 357–377.

Harmer SL, Hogenesch JB, Straume M, Chang HS, Han B, Zhu T, Wang X, Kreps JA, Kay SA (2000) Orchestrated transcription of key pathways in *Arabidopsis* by the circadian clock. Science 290: 2110–2113.

Harmer SL, Panda S, Kay SA (2001) Molecular bases of circadian rhythms. Annu. Rev. Cell Dev. Biol. 17: 215–254.

Hastings JW, Sweeney BM (1957) On the mechanism of temperature compensation in a biological clock. Proc. Natl. Acad. Sci. USA 43: 804–811.

Hazen SP, Schultz TF, Pruneda-Paz JL, Borevitz JO, Ecker JR, Kay SA (2005) *LUX ARRHYTHMO* encodes a Myb domain protein essential for circadian rhythms. Proc. Natl. Acad. Sci. USA 102: 10387–10392.

Helfer A, Nusinow DA, Chow BY, Gehrke AR, Bulyk ML, Kay SA (2011) *LUX ARRHYTHMO* encodes a nighttime repressor of circadian gene expression in the *Arabidopsis* core clock. Curr. Biol. 21: 126–133.

Herrero E, Kolmos E, Bujdoso N, Yuan Y, Wang M, Berns MC, Uhlworm H, Coupland G, Saini R, Jaskolski M, Webb A, Gonçalves J, Davis SJ (2012) EARLY FLOWERING4 recruitment of EARLY FLOWERING3 in the nucleus sustains the *Arabidopsis* circadian clock. Plant Cell 24: 428–443.

Hicks KA, Millar AJ, Carré IA, Somers DE, Straume M, Meeks-Wagner DR, Kay SA (1996a) Conditional circadian dysfunction of the *Arabidopsis early-flowering 3* mutant. Science 274: 790–792.

Hicks KA, Sundas A, Meeks-Wagner DR (1996b) *Arabidopsis* early-flowering mutants reveal multiple levels of regulation in the vegetative-to-floral transition. Semin. Cell Dev. Biol. 7: 409–418.

Hong S, Song H-R, Lutz K, Kerstetter RA, Michael TP, McClung CR (2010) Type II protein arginine methyltransferase PRMT5 is required for circadian period determination in *Arabidopsis thaliana*. Proc. Natl. Acad. Sci. USA 107: 21211–21216.

Hoskins AA, Moore MJ (2012) The spliceosome: A flexible, reversible macromolecular machine. Trends Biochem. Sci. 37: 179–188.

Huang ZJ, Curtin KD, Rosbash M (1995) PER protein interactions and temperature compensation of a circadian clock in *Drosophila*. Science 267: 1169–1172.

Huang W, Pérez-García P, Pokhilko A, Millar AJ, Antoshechkin I, Riechmann JL, Mas P (2012) Mapping the core of the *Arabidopsis* circadian clock defines the network structure of the oscillator. Science 336: 75–79.

Huq E, Tepperman JM, Quail PH (2000) GIGANTEA is a nuclear protein involved in phytochrome signaling in *Arabidopsis*. Proc. Natl. Acad. Sci. USA 97: 9654–9658.

Ibanez C, Ramos A, Acebo P, Contreras A, Casado R, Allona I, Aragoncillo C (2008) Overall alteration of circadian clock gene expression in the chestnut cold response. PLoS One 3: e3567.

Jaglo-Ottosen KR, Gilmour SJ, Zarka DG, Schabenberger O, Thomashow MF (1998) Arabidopsis CBF1 overexpression induces COR genes and enhances freezing tolerance. Science 280: 104–106.

James AB, Monreal JA, Nimmo GA, Kelly CL, Herzyk P, Jenkins GI, Nimmo HG (2008) The circadian clock in *Arabidopsis* roots is a simplified slave version of the clock in shoots. Science 322: 1832–1835.

James AB, Syed NH, Bordage S, Marshall J, Nimmo GA, Jenkins GI, Herzy P, Brown JWS, Nimmo HG (2012) Alternative splicing mediates responses of the *Arabidopsis* circadian clock to temperature changes. Plant Cell 24: 961–981.

Jones MA, Covington MF, DiTacchio L, Vollmers C, Panda S, Harmer SL (2010) Jumonji domain protein JMJD5 functions in both the plant and human circadian systems. Proc. Natl. Acad. Sci. USA 107: 21623–21628.

Kaplan F, Kopka J, Haskell DW, Zhao W, Schiller KC, Gatzke N, Sung DY, Guy CL (2004) Exploring the temperature-stress metabolome of *Arabidopsis*. Plant Physiol. 136: 4159–4168.

Karlsson HG, Johnsson A (1972) A feedback model for biological rhythms. II. Comparisons with experimental results, especially on the petal rhythm of Kalanchoe. J. Theor. Biol. 36: 175–194.

Kikis EA, Khanna R, Quail PH (2005) ELF4 is a phytochrome-regulated component of a negative-feedback loop involving the central oscillator components CCA1 and LHY. Plant J. 44: 300–313.

Kim W-Y, Fujiwara S, Suh S-S, Kim J, Kim Y, Han L, David K, Putterill J, Nam HG, Somers DE (2007) ZEITLUPE is a circadian photoreceptor stabilized by GIGANTEA in blue light. Nature 449: 356–360.

Kim DH, Doyle MR, Sung S, Amasino RM (2009) Vernalization: Winter and the timing of flowering in plants. Annu. Rev. Cell Dev. Biol. 25: 277–299.

Knight H, Trewavas AJ, Knight MR (1996) Cold calcium signaling in *Arabidopsis* involves two cellular pools and a change in calcium signature after acclimation. Plant Cell 8: 489–503.

Kolmos E, Nowak M, Werner M, Fischer K, Schwarz G, Mathews S, Schoof H, Nagy F, Bujnicki JM, Davis SJ (2009) Integrating ELF4 into the circadian system through combined structural and functional studies. HFSP J. 3: 350–366.

Koornneef M, Hanhart CJ, van der Veen JH (1991) A genetic and physiological analysis of late flowering mutants in *Arabidopsis thaliana*. Mol. Gen. Genet. 229: 57–66.

Kumar SV, Wigge PA (2010) H2A.Z-containing nucleosomes mediate the thermosensory response in *Arabidopsis*. Cell 140: 136–147.

Li G, Siddiqui H, Teng Y, Lin R, Wan X-Y, Li J, Lau O-S, Ouyang X, Dai M, Wan J, Devlin PF, Deng XW, Wang H (2011) Coordinated transcriptional regulation underlying the circadian clock in *Arabidopsis*. Nat. Cell Biol. 13: 616–622.

Liu Y, Garceau NY, Loros JJ, Dunlap JC (1997) Thermally regulated translational control of FRQ mediates aspects of temperature responses in the Neurospora circadian clock. Cell 89: 477–486.

Liu Q, Kasuga M, Sakuma Y, Abe H, Miura S, Yamaguchi-Shinozaki K, Shinozaki K (1998) Two transcription factors, DREB1 and DREB2, with an EREBP/AP2 DNA binding domain separate two cellular signal transduction pathways in drought- and low-temperature-response gene expression, respectively in *Arabidopsis*. Plant Cell 10: 1391–1406.

Locke JCW, Southern MM, Kozma-Bognar L, Hibberd V, Brown PE, Turner MS, Millar AJ (2005) Extension of a genetic network model by iterative experimentation and mathematical analysis. Mol. Syst. Biol. 1: 0013.

Lou P, Xie Q, Xu X, Edwards CE, Brock MT, Weinig C, McClung CR (2011) Genetic architecture of the circadian clock and flowering time in *Brassica rapa*. Theor. Appl. Genet. 123: 397–409.

Love J, Dodd AN, Webb AAR (2004) Circadian and diurnal calcium oscillations encode photoperiodic information in *Arabidopsis*. Plant Cell 16: 956–966.

Lu SX, Knowles SM, Webb CJ, Celaya RB, Cha C, Siu JP, Tobin EM (2011) The Jumonji C domain-containing protein JMJ30 regulates period length in the *Arabidopsis* circadian clock. Plant Physiol. 155: 906–915.

de Mairan J (1729) Observation botanique. Hist. Acad. Roy. Sci.: 35–36.

Majercak J, Sidote D, Hardin PE, Edery I (1999) How a circadian clock adapts to seasonal decreases in temperature and day length. Neuron 24: 219–230.

Martin-Tryon EL, Kreps JA, Harmer SL (2007) GIGANTEA acts in blue light signaling and has biochemically separable roles in circadian clock and flowering time regulation. Plant Physiol. 143: 473–486.

Martinière A, Shvedunova M, Thomson AJW, Evans NH, Penfield S, Runions J, McWatters HG (2011) Homeostasis of plasma membrane viscosity in fluctuating temperatures. New Phytol. 192: 328–337.

Martino-Catt S, Ort DR (1992) Low temperature interrupts circadian regulation of transcriptional activity in chilling-sensitive plants. Proc. Natl. Acad. Sci. USA 89: 3731–3735.

Más P (2008) Circadian clock function in *Arabidopsis thaliana*: Time beyond transcription. Trends Cell Biol. 18: 273–281.

Más P, Alabadí D, Yanovsky MJ, Oyama T, Kay SA (2003) Dual role of TOC1 in the control of circadian and photomorphogenic responses in *Arabidopsis*. Plant Cell 15: 223–236.

Matsumoto A, Tomioka K, Chiba Y, Tanimura T (1999) *tim[rit]* lengthens circadian period in a temperature-dependent manner through suppression of PERIOD protein cycling and nuclear localization. Mol. Cell. Biol. 19: 4343–4354.

Matsushika A, Makino S, Kojima M, Mizuno T (2000) Circadian waves of expression of the APRR1/TOC1 family of pseudo-response regulators in *Arabidopsis thaliana*: Insight into the plant circadian clock. Plant Cell Physiol. 41: 1002–1012.

McAinsh MR, Webb AAR, Taylor JE, Hetherington AM (1995) Stimulus-induced oscillations in guard cell cytosolic free calcium. Plant Cell 7: 1207–1219.

McClung CR, Davis SJ (2010) Ambient thermometers in plants: From physiological outputs towards mechanisms of thermal sensing. Curr. Biol. 20: R1086–R1092.

McWatters HG, Bastow RM, Hall A, Millar AJ (2000) The *ELF3 zeitnehmer* regulates light signalling to the circadian clock. Nature 408: 716–720.

McWatters HG, Kolmos E, Hall A, Doyle MR, Amasino RM, Gyula P, Nagy F, Millar AJ, Davis SJ (2007) ELF4 is required for oscillatory properties of the circadian clock. Plant Physiol. 144: 391–401.

Mehra A, Shi M, Baker CL, Colot HV, Loros JJ, Dunlap JC (2009) A role for casein kinase 2 in the mechanism underlying circadian temperature compensation. Cell 137: 749–760.

Michael TP, Salomé PA, McClung CR (2003a) Two *Arabidopsis* circadian oscillators can be distinguished by differential temperature sensitivity. Proc. Natl. Acad. Sci. USA 100: 6878–6883.

Michael TP, Salomé PA, Yu HJ, Spencer TR, Sharp EL, Alonso JM, Ecker JR, McClung CR (2003b) Enhanced fitness conferred by naturally occurring variation in the circadian clock. Science 302: 1049–1053.

Michaels SD, Amasino RM (1999) *FLOWERING LOCUS C* encodes a novel MADS domain protein that acts as a repressor of flowering. Plant Cell 11: 949–956.

Mizoguchi T, Wright L, Fujiwara S, Cremer F, Lee K, Onouchi H, Mouradov A, Fowler S, Kamada H, Putterill J, Coupland G (2005) Distinct roles of *GIGANTEA* in promoting flowering and regulating circadian rhythms in *Arabidopsis*. Plant Cell 17: 2255–2270.

Mizuno T (2004) Plant response regulators implicated in signal transduction and circadian rhythm. Curr. Opin. Plant Biol. 7: 499–505.

de Montaigu A, Tóth R, Coupland G (2010) Plant development goes like clockwork. Trends Genet. 26: 296–306.

Nakamichi N, Kita M, Ito S, Sato E, Yamashino T, Mizuno T (2005) PSEUDO-RESPONSE REGULATORS, PRR9, PRR7 and PRR5, together play essential roles close to the circadian clock of *Arabidopsis thaliana*. Plant Cell Physiol. 46: 686–698.

Nakamichi N, Kusano M, Fukushima A, Kita M, Ito S, Yamashino T, Saito K, Sakakibara H, Mizuno T (2009) Transcript profiling of an *Arabidopsis* PSEUDO RESPONSE REGULATOR arrhythmic triple mutant reveals a role for the circadian clock in cold stress response. Plant Cell Physiol. 50: 447–462.

Nakamichi N, Kiba T, Henriques R, Mizuno T, Chua N-H, Sakakibara H (2010) PSEUDO-RESPONSE REGULATORS 9, 7 and 5 are transcriptional repressors in the *Arabidopsis* circadian clock. Plant Cell 22: 594–605.

Niwa Y, Yamashino T, Mizuno T (2009) The circadian clock regulates the photoperiodic response of hypocotyl elongation through a coincidence mechanism in *Arabidopsis thaliana*. Plant Cell Physiol. 50: 838–854.

Njus D, Sulzman FM, Hastings JW (1974) Membrane model for the circadian clock. Nature 248: 116–120.

Nozue K, Maloof JN (2006) Diurnal regulation of plant growth. Plant Cell Environ. 29: 396–408.

Nozue K, Covington MF, Duek PD, Lorrain S, Fankhauser C, Harmer SL, Maloof JN (2007) Rhythmic growth explained by coincidence between internal and external cues. Nature 448: 358–361.

Nusinow DA, Helfer A, Hamilton EE, King JJ, Imaizumi T, Schultz TF, Farre EM, Kay SA (2011) The ELF4-ELF3-LUX complex links the circadian clock to diurnal control of hypocotyl growth. Nature 475: 398–402.

Ouyang Y, Andersson CR, Kondo T, Golden SS, Johnson CH (1998) Resonating circadian clocks enhance fitness in cyanobacteria. Proc. Natl. Acad. Sci. USA 95: 8660–8664.

Park DH, Somers DE, Kim YS, Choy YH, Lim HK, Soh MS, Kim HJ, Kay SA, Nam HG (1999) Control of circadian rhythms and photoperiodic flowering by the *Arabidopsis GIGANTEA* gene. Science 285: 1579–1582.

Pennycooke JC, Cheng H, Roberts SM, Yang Q, Rhee SY, Stockinger EJ (2008) The low temperature-responsive, *Solanum CBF1* genes maintain high identity in their upstream regions in a genomic environment undergoing gene duplications, deletions, and rearrangements. Plant Mol. Biol. 67: 483–497.

Perales M, Más P (2007) A functional link between rhythmic changes in chromatin structure and the *Arabidopsis* biological clock. Plant Cell 19: 2111–2123.

Perales M, Portolés S, Más P (2006) The proteasome-dependent degradation of CKB4 is regulated by the *Arabidopsis* biological clock. Plant J. 46: 849–860.

Pittendrigh CS (1954) On the temperature independence in the clock system controlling emergence time in *Drosophila*. Proc. Natl. Acad. Sci. USA 40: 1018–1029.

Pittendrigh CS, Bruce VG (1959) Daily rhythms as coupled oscillator systems and their relation to thermoperiodism and photoperiodism. In: Withrow RB (ed). Photoperiodism and related phenomena in plants and animals. American Association for the Advancement of Science, Washington, DC, pp. 475–505.

Pittendrigh C, Bruce V, Kaus P (1958) On the significance of transients in daily rhythms. Proc. Natl. Acad. Sci. USA 44: 965–973.

Pokhilko A, Fernández AP, Edwards KD, Southern MM, Halliday KJ, Millar AJ (2012) The clock gene circuit in *Arabidopsis* includes a repressilator with additional feedback loops. Mol. Syst. Biol. 8: 574.

Portolés S, Más P (2010) The functional interplay between protein kinase CK2 and CCA1 transcriptional activity is essential for clock temperature compensation in *Arabidopsis*. PLoS Genet. 6: e1001201.

Pruneda-Paz JL, Kay SA (2010) An expanding universe of circadian networks in higher plants. Trends Plant Sci. 15: 259–265.

Pyl E-T, Piques M, Ivakov A, Schulze W, Ishihara H, Stitt M, Sulpice R (2012) Metabolism and growth in *Arabidopsis* depend on the daytime temperature but are temperature-compensated against cool nights. Plant Cell 24: 2443–2469.

Ramos A, Perez-Solis E, Ibanez C, Casado R, Collada C, Gomez L, Aragoncillo C, Allona I (2005) Winter disruption of the circadian clock in chestnut. Proc. Natl. Acad. Sci. USA 102: 7037–7042.

Rawat R, Takahashi N, Hsu PY, Jones MA, Schwartz J, Salemi MR, Phinney BS, Harmer SL (2011) REVEILLE8 and PSEUDO-REPONSE REGULATOR5 form a negative feedback loop within the *Arabidopsis* circadian clock. PLoS Genet. 7: 16.

Rédei GP (1962) Supervital mutants of *Arabidopsis*. Genetics 47: 443–460.

Resco V, Hartwell J, Hall A (2009) Ecological implications of plants' ability to tell the time. Ecol. Lett. 12: 583–592.

Rohde A, Bhalerao RP (2007) Plant dormancy in the perennial context. Trends Plant Sci. 12: 217–223.

Ruoff P (1992) Introducing temperature-compensation in any reaction kinetic oscillator model. J. Interdiscipl. Cycle Res. 23: 92–99.

Sahar S, Sassone-Corsi P (2012) Regulation of metabolism: The circadian clock dictates the time. Trends Endocrinol. Metab. 23: 1–8.

Salathia N, Davis SJ, Lynn JR, Michaels SD, Amasino RM, Millar AJ (2006) FLOWERING LOCUS C-dependent and -independent regulation of the circadian clock by the autonomous and vernalization pathways. BMC Plant Biol. 6: 10.

Salomé PA, McClung CR (2005) *PSEUDO-RESPONSE REGULATOR 7 and 9* are partially redundant genes essential for the temperature responsiveness of the *Arabidopsis* circadian clock. Plant Cell 17: 791–803.

Salomé PA, Xie Q, McClung CR (2008) Circadian timekeeping during early *Arabidopsis* development. Plant Physiol. 147: 1110–1125.

Salomé PA, Weigel D, McClung CR (2010) The role of the *Arabidopsis* morning loop components CCA1, LHY, PRR7 and PRR9 in temperature compensation. Plant Cell 22: 3650–3661.

Sanchez SE, Petrillo E, Beckwith EJ, Zhang X, Rugnone ML, Hernando CE, Cuevas JC, Godoy Herz MA, Depetris-Chauvin A, Simpson CG, Brown JWS, Cerdán PD, Borevitz JO, Mas P, Ceriani MF, Kornblihtt AR, Yanovsky MJ (2010) A methyl transferase links the circadian clock to the regulation of alternative splicing. Nature 468: 112–116.

Sawa M, Kay SA (2011) GIGANTEA directly activates flowering locus T in *Arabidopsis thaliana*. Proc. Natl. Acad. Sci. USA 108: 11698–11703.

Sawa M, Nusinow DA, Kay SA, Imaizumi T (2007) FKF1 and GIGANTEA complex formation is required for day-length measurement in *Arabidopsis*. Science 318: 261–265.

Schaffer R, Ramsay N, Samach A, Corden S, Putterill J, Carré IA, Coupland G (1998) The *late elongated hypocotyl* mutation of *Arabidopsis* disrupts circadian rhythms and the photoperiodic control of flowering. Cell 93: 1219–1229.

Seo PJ, Park M-J, Lim M-H, Kim S-G, Lee M, Baldwin IT, Park C-M (2012) A self-regulatory circuit of CIRCADIAN CLOCK-ASSOCIATED1 underlies the circadian clock regulation of temperature responses in *Arabidopsis*. Plant Cell 24: 2427–2442.

Sheldon CC, Burn JE, Perez PP, Metzger J, Edwards JA, Peacock WJ, Dennis ES (1999) The *FLF* MADS box gene: A repressor of flowering in *Arabidopsis* regulated by vernalization and methylation. Plant Cell 11: 445–458.

Sheldon CC, Finnegan EJ, Peacock WJ, Dennis ES (2009) Mechanisms of gene repression by vernalization in *Arabidopsis*. Plant J. 59: 488–498.

Shell GSG, Lang ARG (1976) Movements of sunflower leaves over a 24-h period. Agric. Meteorol. 16: 161–170.

Shinozaki K, Yamaguchi-Shinozaki K (2000) Molecular responses to dehydration and low temperature: Differences and cross-talk between two stress signaling pathways. Curr. Opin. Plant Biol. 3: 217–223.

Somers DE (2012) The *Arabidopsis* clock: Time for an about-face? Genome Biol. 13: 153.

Somers DE, Fujiwara S (2009) Thinking outside the F-box: Novel ligands for novel receptors. Trends Plant Sci. 14: 206–213.

Somers DE, Webb AAR, Pearson M, Kay SA (1998) The short-period mutant, *toc1-1*, alters circadian clock regulation of multiple outputs throughout development in *Arabidopsis thaliana*. Development 125: 485–494.

Somers DE, Schultz TF, Milnamow M, Kay SA (2000) ZEITLUPE encodes a novel clock-associated PAS protein from *Arabidopsis*. Cell 101: 319–329.

Steponkus PL, Uemura M, Joseph RA, Gilmour SJ, Thomashow MF (1998) Mode of action of the *COR15a* gene on the freezing tolerance of *Arabidopsis thaliana*. Proc. Natl. Acad. Sci. USA 95: 14570–14575.

Stockinger EJ, Gilmour SJ, Thomashow MF (1997) *Arabidopsis thaliana* CBF1 encodes an AP2 domain-containing transcriptional activator that binds to the C-repeat/DRE, a cis-acting DNA regulatory element that stimulates transcription in response to low temperature and water deficit. Proc. Natl. Acad. Sci. USA 94: 1035–1040.

Sugano S, Andronis C, Green RM, Wang Z-Y, Tobin EM (1998) Protein kinase CK2 interacts with and phosphorylates the *Arabidopsis* circadian clock-associated 1 protein. Proc. Natl. Acad. Sci. USA 95: 11020–11025.

Swarup K, Alonso-Blanco C, Lynn JR, Michaels SD, Amasino RM, Koornneef M, Millar AJ (1999) Natural allelic variation identifies new genes in the *Arabidopsis* circadian system. Plant J. 20: 67–77.

Thain SC, Hall A, Millar AJ (2000) Functional independence of multiple circadian clocks that regulate plant gene expression. Curr. Biol. 10: 951–956.

Thines B, Harmon FG (2010) Ambient temperature response establishes ELF3 as a required component of the core *Arabidopsis* circadian clock. Proc. Natl. Acad. Sci. USA 107: 3257–3262.

Thomashow MF (1999) Plant cold acclimation: Freezing tolerance genes and regulatory mechanisms. Annu. Rev. Plant Physiol. Plant Mol. Biol. 50: 571–599.

Trewavas A (1999) Le Calcium, C'est la Vie: Calcium makes waves. Plant Physiol. 120: 1–6.

Vandenbussche F, Verbelen JP, Van Der Straeten D (2005) Of light and length: Regulation of hypocotyl growth in *Arabidopsis*. Bioessays 27: 275–284.

Vogel JT, Zarka DG, Van Buskirk HA, Fowler SG, Thomashow MF (2005) Roles of the CBF2 and ZAT12 transcription factors in configuring the low temperature transcriptome of *Arabidopsis*. Plant J. 41: 195–211.

Wang Z-Y, Tobin EM (1998) Constitutive expression of the *CIRCADIAN CLOCK ASSOCIATED 1* (*CCA1*) gene disrupts circadian rhythms and suppresses its own expression. Cell 93: 1207–1217.

Whitehead K, Pan M, Masumura K-i, Bonneau R, Baliga NS (2009) Diurnally entrained anticipatory behavior in Archaea. PLoS One 4: e5485.

Woelfle MA, Ouyang Y, Phanvijhitsiri K, Johnson CH (2004) The adaptive value of circadian clocks. Curr. Biol. 14: 1481–1486.

Wolkovich EM, Cook BI, Allen JM, Crimmins TM, Betancourt JL, Travers SE, Pau S, Regetz J, Davies TJ, Kraft NJB, Ault TR, Bolmgren K, Mazer SJ, McCabe GJ, McGill BJ, Parmesan C, Salamin N, Schwartz MD, Cleland EE (2012) Warming experiments underpredict plant phenological responses to climate change. Nature 485: 494–497.

Wood NT, Haley A, Viry-Moussaïd M, Johnson CH, van der Luit AH, Trewavas AJ (2001) The calcium rhythms of different cell types oscillate with different circadian phases. Plant Physiol. 125: 787–796.

Xu X, Hotta CT, Dodd AN, Love J, Sharrock R, Lee YW, Xie Q, Johnson CH, Webb AAR (2007) Distinct light and clock modulation of cytosolic free Ca^{2+} oscillations and rhythmic *CHLOROPHYLL A/B BINDING PROTEIN2* promoter activity in *Arabidopsis*. Plant Cell 19: 3474–3490.

Yakir E, Hilman D, Harir Y, Green RM (2007) Regulation of output from the plant circadian clock. FEBS J. 274: 335–345.

Yamashino T, Ito S, Niwa Y, Kunihiro A, Nakamichi N, Mizuno T (2008) Involvement of *Arabidopsis* clock-associated pseudo-response regulators in diurnal oscillations of gene expression in the presence of environmental time cues. Plant Cell Physiol. 49: 1839–1850.

Yerushalmi S, Green RM (2009) Evidence for the adaptive significance of circadian rhythms. Ecol. Lett. 12: 970–981.

Zagotta MT, Shannon S, Jacobs C, Meeks-Wagner DR (1992) Early-flowering mutants of *Arabidopsis thaliana*. Aust. J. Plant Physiol. 19: 411–418.

Zagotta MT, Hicks KA, Jacobs CI, Young JC, Hangarter RP, Meeks-Wagner DR (1996) The *Arabidopsis* ELF3 gene regulates vegetative photomorphogenesis and the photoperiodic induction of flowering. Plant J. 10: 691–702.

Zhang EE, Kay SA (2010) Clocks not winding down: Unravelling circadian networks. Nat. Rev. Mol. Cell Biol. 11: 764–776.

7 Temperature and plant immunity

Jian Hua

7.1 Introduction

Plants have evolved effective immune systems and consequently are often not diseased in spite of coexisting with a myriad of pathogens. Immune responses can, however, be energetically costly and have adverse effects on plant growth and development. Plants therefore balance immune responses and plant growth to achieve maximum fitness. Emerging evidence indicates that plants utilize environmental signals to anticipate and prepare for incoming abiotic and biotic stresses. As pathogen virulence is modulated by environmental factors including temperature, it is conceivable that plants use temperature information to enhance immunity when it is most needed, thereby minimizing fitness costs.

It has long been noted that temperature influences disease resistance to a variety of pathogenic bacteria, fungi, virus, and insects. However, temperature modulation of plant–pathogen interaction appears to be more difficult to understand than the temperature modulation of other processes. The effects of temperature on plant immunity can be compounded by the effects of other environmental variables such as light and humidity, which are often associated with temperature in the field. In addition, the outcome of susceptibility or resistance to a pathogen is determined not only by plant immunity but also by pathogen virulence, both of which display diverse strategies and environmental regulation. It is therefore not too surprising that variable and sometimes opposite temperature effects have been observed for plant disease resistance. Advances in deciphering the molecular mechanisms underlying plant immunity and pathogen virulence have facilitated understanding of temperature sensitivity in these processes. Rather than being an indirect effect resulting from changes in overall cell physiology, temperature has been found to have a direct effect on plant immunity. Putative temperature-sensitive components have been identified in plant defense responses in addition to factors influencing the temperature sensitivity of immunity. Given the impact of current climate change on

Temperature and Plant Development, First Edition. Edited by Keara A. Franklin and Philip A. Wigge.

plant disease resistance (Garrett et al. 2006; Gaudet et al. 2011), there is even greater need to understand how temperature variations alter interactions between plants and pathogens.

This chapter aims to analyze common features of the temperature effects on plant immunity and present current knowledge of molecular basis of temperature sensitivity in this process. Mechanistic study of the environmental control of plant immunity is still in its infancy, and a further investigation of this regulation will increase basic knowledge of adaptation in plants, thereby contributing to sustainable agriculture and ecosystems.

7.2 Plant immunity

Plant immunity is constantly evolving, due to an 'arms race' between pathogens and plants at both the individual and population levels. As pathogens use a variety of strategies for infection, invasion, feeding, and propagation, plants employ a variety of defense mechanisms responding to both shared and unique pathogen features (Dodds and Rathjen 2010). Vast advances have been made in the fields of plant immunity and pathogen virulence over the past fifteen years. In the following text is a brief summary of current knowledge.

7.2.1 Immunity against microbial pathogens

Resistance against microbial pathogens consists of multiple layers, mainly involving a general species resistance (also named basal resistance) and a cultivar-specific resistance (Nurnberger et al. 2004; Chisholm et al. 2006; Jones and Dangl 2006). The existence of multiple layers of defense responses likely arises from the coevolution between plants and pathogens. The first layer of basal resistance includes constitutive protective barriers (such as waxes and cell walls) and inducible reactions to common features of pathogens such as 'pathogen-associated molecular patterns' (PAMPs). In the PAMP-triggered immunity (PTI), plants use 'pattern recognition receptors' (PRRs) to perceive shared pathogen molecules (Boller and Felix 2009; Monaghan and Zipfel 2012). All known PRRs are transmembrane proteins containing ligand-binding ectodomains. For instance, the bacterial PAMPs, flagellin and EF-Tu, are recognized by the plant PRRs FLAGELLIN SENSING 2 (FLS2) and EF-TU RECEPTOR (EFR), respectively, both of which are leucine-rich repeat (LRR) receptor kinases. These receptors act with other kinases such as BRASSINOSTEROID INSENSITIVE1-ASSOCIATED KINASE1 (BAK1) in protein complexes to transduce signals across the plasma membrane. The activation of PRRs is followed by series of events including very early ion fluxes, oxidative burst, and mitogen-activated protein kinase (MAPK) activation; early responses including ethylene biosynthesis, receptor endocytosis, and gene activation; and late responses such as callose deposition (Boller and Felix 2009). Together, these events in PTI effectively block the invasion of most pathogens.

Pathogens have consequently evolved effectors to target core components of PTI to bypass the first layer of plant defense (Deslandes and Rivas 2012; Rafiqi et al. 2012). Plants, in turn, evolved another layer of immunity to recognize effector proteins and mount a faster and more robust defense response than that observed in PTI. Cultivar-specific resistance or effector-triggered immunity (ETI) is induced upon a specific recognition of the pathogen effector gene by the disease resistance (*R*) gene of the host plant. This 'gene-for-gene' interaction (Flor 1971) leads to rapid and efficient defense responses, including a form of programmed cell death named the hypersensitive response (HR) to restrict the growth of biotrophic pathogens (Hammond-Kosack and Jones 1996). HR is characterized by rapid calcium ion fluxes, an extracellular oxidative burst, and transcriptional reprogramming (Scheel 1998; Greenberg and Yao 2004). *R* gene activation in ETI also induces systemic acquired resistance (SAR) in distal regions to prepare plants for subsequent pathogen attacks. Salicylic acid (SA) is a central regulator of SAR (Durrant and Dong 2004; Vlot et al. 2009).

Over 40 *R* genes have been cloned in various plant species and mostly encode five classes of proteins (Dangl and Jones 2001; Martin et al. 2003). The largest R protein class is the nucleotide-binding (NB) site–LRR (NB-LRR) family (Caplan et al. 2008; Lukasik and Takken 2009). NB-LRR proteins are structurally similar to the animal NOD-like receptor family proteins responsible for pathogen recognition in animal immunity (Ausubel 2005). There are approximately 150 *NB-LRR* protein-encoding genes in *Arabidopsis thaliana* which are highly divergent among different accessions (Meyers et al. 2003). The *NB-LRR* genes often reside in gene clusters, perhaps facilitating gene recombination to generate new *R* genes for novel pathogen recognition. The N-termini of NB-LRR proteins are either of the Toll and interleukin-1 receptor (TIR) type or the *coiled-coil* (CC) type. With a few exceptions, the proteins ENHANCED DISEASE SUSCEPTIBILITY 1 (EDS1) and PHYTOALEXIN-DEFICIENT 4 (PAD4) are required for the function of TIR-NB-LRR proteins while NON-RACE-SPECIFIC DISEASE RESISTANCE 1 (NDR1) is required for CC-NB-LRR proteins (Glazebrook 2001; Wiermer et al. 2005). Recently, RPS4 and RPS6, R proteins against bacterial pathogens, are each found to form protein complexes with EDS1, and these interactions can be disrupted by bacterial effector proteins (Bhattacharjee et al. 2011; Heidrich et al. 2011).

R proteins are central to pathogen recognition and signaling (Rafiqi et al. 2009; Elmore et al. 2011; Heidrich et al. 2012). They are activated upon direct binding to pathogen effector proteins or indirect recognition of plant proteins modified by effectors. Intramolecular interactions among the N-terminus, NB, and LRR domains are crucial for R protein activity, and a number of chaperone proteins including HEAT SHOCK PROTEIN 90 (HSP90), SUPPRESSOR OF G2 ALLELE OF *skp1* (SGT1), and HEAT SHOCK COGNATE 70 (HSC70) are required for the maturation and/or the activation of R proteins. NB-LRR proteins have diverse cellular localization and form distinct complexes possibly leading to diverse immune outputs.

Nuclear accumulation is essential for the immune activities of a number of NB-LRR proteins. It has been observed that effectors can induce the nuclear accumulation of NB-LRR MLA10 and N proteins that confer resistance to powdery mildew in barley and tobacco mosaic virus (TMV) in tobacco, respectively (Burch-Smith et al. 2007; Shen et al. 2007). Nuclear accumulation has also been shown to be critical for the function of two NB-LRR proteins, SNC1 and RPS4, that confer resistance against bacterial pathogens (Wirthmueller et al. 2007; Monaghan et al. 2010). R proteins may directly regulate gene expression in the nucleus. The barley MLA10 R protein interacts with WRKY transcriptional repressors of PTI, thereby interfering with their function (Shen et al. 2007). The SNC1 R protein has been shown to associate with the transcriptional corepressor TOPLESS-RELATED 1 (TPR1) which functions to repress the expression of negative regulators of immune responses (Zhu et al. 2010b).

7.2.2 *Immunity against necrotrophic pathogens*

Cell death triggered by R proteins is effective against biotrophic or hemibiotrophic pathogens but not necrotrophic pathogens which benefit from host cell death. In fact, effectors identified from necrotrophic pathogens have been shown to target R proteins to trigger cell death and thus confer disease to plants (Oliver and Solomon 2010). Immunity against necrotrophic pathogens utilizes ethylene and jasmonate (JA) rather than SA for local and systemic resistance (Glazebrook 2005). Although JA and SA were thought to be antagonistic, a recent study has shown fine spatial and temporal regulation between SA-induced defense against biotrophic pathogens and JA-mediated resistance against necrotrophic pathogens (Spoel et al. 2007).

7.2.3 *Immunity against herbivorous insects*

Plants perceive the invasion of herbivorous insects through the recognition of their oral secretions and damaged plant cells. The perception of herbivore invasion is transduced by calcium ion fluxes, protein phosphorylation, and the JA pathway to induce production of defense compounds that target insect physiological processes and/or volatiles that attract insect predators (Howe and Jander 2008). Despite their different lifestyles, the interactions between pathogenic insects with plants and microbes with plants share some similarities. Effector proteins identified from the saliva of insects were shown, like effectors from microbial pathogens, to target plant defense response components (Hogenhout and Bos 2011). Plants also use *R* genes to detect insect invasion, and at least three *R* genes coding for NB-LRR proteins have been identified to confer insect resistance (Hogenhout and Bos 2011).

7.2.4 Immunity against viruses

Plants use two major immune strategies to defend against viruses. One is *R* gene-mediated defense responses resulting in a local HR and a long-distance SAR. About ten *R* genes against viruses have been cloned in plants; and their encoded R proteins often recognize viral coat proteins (Soosaar et al. 2005). The second strategy used against virus invasion is RNA silencing (Ding 2010; Wang et al. 2012). As the viral genome is replicated within plant cells, the unique features of viral invasive nucleic acids become a nonself recognition target for plants (Alvarado and Scholthof 2009). Viral small interference RNAs (siRNAs) are generated by dicer-like (DCL) proteins from double-stranded RNA, whose formation and amplification are facilitated by RNA-dependent RNA polymerase (RDR). Accumulation of siRNAs triggers RNA silencing in plants, leading to viral RNA degradation or viral DNA methylation. Viruses have therefore evolved various strategies such as expressing viral suppressors of RNA silencing to overcome this antiviral mechanism (Baulcombe 2002; Silhavy and Burgyan 2004).

In summary, plant immunity has coevolved with pathogens. Defense responses are initiated by the perception of both pathogens and pathogen-induced changes. A diverse array of detection strategies coupled with signaling amplification pathways makes plant immunity a versatile and effective defense system.

7.3 Temperature effects on plant disease resistance

Susceptibility or resistance to pathogens at a particular temperature is dependent on the combined effects of temperature on pathogen virulence and plant immunity. Changes in temperature can inhibit or induce disease resistance with varying thresholds. Although temperature effects appear different or even contrasting among diverse interactions, a careful examination of the relevant reports, as summarized in the following text, reveals two major categories of immunity modulation by temperature: high-temperature-mediated suppression and low-temperature-mediated suppression. The former is mostly mediated by *R* genes and the latter by RNA silencing. In other specific cases, low or high temperatures can also induce immunity.

7.3.1 High-temperature suppression of disease resistance

High temperature often inhibits plant disease resistance (Dropkin 1969; Wang et al. 2009). A closer inspection of these cases indicates that they are associated with *R* gene-mediated resistance. The classical example is the TIR-NB-LRR-encoding *N* gene against TMV, which is inactivated at temperatures above 28°C (Samuel 1931). This property of temperature sensitivity is retained when the *N* gene is transferred in tomato (Whitham et al. 1996). Similarly, the TIR-NB-LRR-coding *Mi-1* gene only confers resistance to root-knot nematodes in tomato at temperatures below 28°C (Hwang et al. 2000). Cf-4 and Cf-9 are tomato *R* genes against the fungal pathogen *Cladosporium fulvum*. The HR mediated by these genes can be suppressed at 33°C (de Jong et al. 2002). In *Arabidopsis*, the *RESISTANCE TO POWDERY MILDEW 8* (*RPW8*) gene encodes a CC-containing R protein against powdery mildew, which induces cell death and defense responses at 22°C but not at temperatures above 30°C (Xiao et al. 2003). Heat sensitivity has been found in a number of resistances against the hemibiotrophic pathogen *Pseudomonas syringae* in *Arabidopsis*. Resistances triggered by effectors AvrRpt2, AvrRps4, and AvrRpm1 at 22°C are inhibited to different extents at 28°C, indicating that immunity mediated by their corresponding *R* genes *RESISTANCE TO PSEUDOMONAS SYRINGAE 2* (*RPS2*), *RESISTANCE TO PSEUDOMONAS SYRINGAE 4* (*RPS4*), and *RPM1* is temperature sensitive (Wang et al. 2009). In addition to *RPS4*, the TIR-NB-LRR type of *R* gene, *SNC1*, also contributes to resistance against AvrRps4 in a temperature-dependent manner (Kim et al. 2010).

A number of autoimmune mutants with upregulated defense responses are also heat sensitive. The *bonzai 1* (*bon1*) mutant of *Arabidopsis* exhibits a dwarf phenotype at 22°C but not at 28°C, resulting from a suppression of defense responses mediated by the *R* gene *SNC1* at elevated temperature (Hua et al. 2001; Yang and Hua 2004). The gain-of-function SNC1 mutant, *snc1-1*, exhibits constitutive defense responses and a dwarf phenotype at 22°C but not at 28°C (Zhang et al. 2003; Yang and Hua 2004). Missense mutations in the NB-LRR *R* gene *RPP4* and a TIR-NB-LRR-LIM *R*-like gene confer growth defects resulting from autoimmune responses below 16°C but not at 22°C (Huang et al. 2010; Yang et al. 2010). Mutations in MAPK/ERK Kinase Kinase 1 (MEKK1) of the MAP kinase cascade lead to autoimmune phenotypes at 22°C which can be suppressed by increasing temperature to 28°C (Ichimura et al. 2006). This temperature sensitivity is associated with *R* gene-mediated resistance as the NB-LRR gene SUPPRESSOR OF mkk1 mkk2 (*SUMM*) is found to mediate the phenotypes of *mekk1* (Zhang et al. 2012b). The loss of autoimmune mutant phenotypes at high temperature is consistent with the notion that *R*-mediated disease resistance is inhibited by heat.

Hybrid necrosis, a post-zygotic incompatibility between inter- or intraspecies, also displays heat sensitivity (Bomblies et al. 2007; Alcazar et al. 2009). Crosses between *Arabidopsis* accessions revealed that 2% of intraspecific crosses exhibited necrosis or cell death. In most cases, the necrosis phenotype was more severe at 16°C than at 23°C. Epistatic interactions consisting of NB-LRR genes are found to be responsible for the hybrid necrosis phenotype. Heat sensitivity of hybrid necrosis therefore likely results from high-temperature suppression of defense responses triggered by *R* genes.

7.3.2 Low-temperature inhibition of plant immunity

Low temperatures are also reported to reduce plant defenses. In these cases, defense responses mainly involve RNA silencing. Tobacco plants infected with the *Cymbidium* ring spot RNA virus were shown to accumulate high levels of virus-derived siRNA that triggers silencing at 27°C but not at 15°C. As a consequence, plants displayed resistance to this virus at high but not low temperatures (Szittya et al. 2003). Similar temperature effects were observed for the DNA virus geminivirus. This was shown to trigger RNA silencing with a higher accumulation of siRNA at 30°C than 25°C (Chellappan et al. 2005). Disruption of the key RNA silencing key component RNA-dependent RNA polymerase 6 (RDR6) renders *Nicotiana benthamiana* more susceptible to several RNA viruses at 27°C than at 21°C, indicating that virus silencing is more active at high temperature (Qu et al. 2005). In a separate study, silencing against geminivirus was found to be greater in plants grown in cycles of 30/26°C (day/night) than 22/18°C (Tuttle et al. 2008).

7.3.3 Disease resistance induced by high and low temperatures

Exceptions exist to the two categories mentioned in Sections 7.3.1 and 7.3.2 of temperature-regulated plant immunity. Exposure to cold (<8°C) for several hours has been shown to reduce infection and colony expansion of powdery mildew in grapevines (Moyer et al. 2010). In *Arabidopsis*, SA (both free and conjugated) accumulates to a higher level at 5°C than at 23°C (Scott et al. 2004). Whether or not cold-induced disease resistance in grapevine is associated with cold-induced SA has yet to be investigated.

Disease resistance can also be induced by relatively high temperatures. The wheat *Yr36* gene encoding a kinase-START protein confers resistance to a broad spectrum of stripe rust races at high temperatures (from 25°C to 35°C) (Fu et al. 2009). An increase of *Yr36* RNA expression at relatively high temperatures may account for the temperature dependency of this resistance.

7.4 The molecular basis for temperature sensitivity in plant immunity

Temperature can affect almost all molecules and their interactions in cells, including RNA, DNA, proteins, and lipids. On the one hand, an organism has the capacity to buffer temperature fluctuations to keep cellular function in homeostasis. On the other hand, it can sense temperature and respond to this signal for a coordinated response. Temperature-sensing mechanisms in plant immunity could include temperature-dependent changes in transcription factor (TF) RNA structure, leading to altered activity controlling virulence gene expression in bacteria (Johansson et al. 2002). A second mechanism could involve temperature-dependent activation of calcium channels, similar to thermosensing mechanisms in animals (Dhaka et al. 2006).

The temperature effects on plant immunity discussed earlier indicate that plant defense responses with similar strategies have similar temperature sensitivities. It is conceivable that temperature, which carries information about daytime and season as well as pathogen pressure, could be sensed by plants to modulate their immune responses for maximal effectiveness against pathogens while minimizing costs on growth. Recent studies are beginning to reveal the molecular basis for temperature perception and sensitivity in plant immunity.

7.4.1 Heat-sensitive NB-LRR R proteins

Molecular genetic studies suggest that the heat sensitivity of NB-LRR proteins may underlie the temperature sensitivity of plant disease resistance. Autoimmune mutants displaying a heat-sensitive phenotype were used as a background for mutant screens to look for heat-stable disease resistance or temperature insensitivity (Zhu et al. 2010a, 2012; Mang et al. 2012). The first insensitive to temperature (*int*) mutant identified resulted from a missense mutation in the LRR region of the *R* gene *SNC1*. This conferred disease resistance at 28°C, a temperature which usually suppresses *SNC1* activity. A cognate mutation introduced into the tobacco *R* gene *N* conferred defense responses at 28°C, a temperature which normally inhibits *N*-mediated resistance. In addition, heat-stable resistance was reverted back to heat-sensitive resistance by additional mutations in the NB domain of

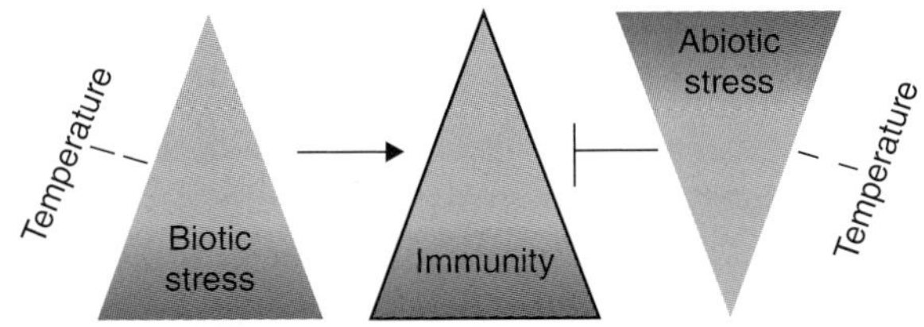

Figure 7.1 Model of high-temperature inhibition of *R*-mediated plant immunity. R proteins recognize pathogen effectors directly or indirectly and undergo conformational changes from inactive to active forms. Some R proteins shuttle to the nucleus after activation to regulate expression of defense genes through their association with transcriptional regulators such as TF. High temperature likely affects the conformation change of R proteins or their interaction with other proteins and thus inhibits their nuclear accumulation. Nuclear accumulation of R proteins is also negatively regulated by the plant hormone ABA that is an indicator of abiotic stresses.

SNC1. It can therefore be concluded that NB-LRR proteins determine temperature sensitivity of the defense response and the temperature sensitivity of *R*-mediated immunity can be altered by a change in the primary sequences of an *R* gene.

The earliest observed effect of temperature on *R* gene activity involves subcellular localization (Zhu et al. 2010a; Mang et al. 2012). Activated R proteins including SNC1, N, and RPS4 were observed to move from the cytosol to the nucleus at 22°C where their function is required for immune responses. Nuclear accumulation was, however, reduced at elevated temperature (28°C). In contrast, the mutant SNC1 protein that confers heat-stable resistance stays in the nucleus even at 28°C. In this way, temperature modulates disease resistance through affecting the nuclear accumulation of R proteins.

How R proteins 'sense' temperature to alter their subcellular localizations is not known. The most likely scenario is that nuclear localization of R proteins requires an active conformation of R and/or its association with particular proteins, which can be disrupted by high temperature (Figure 7.1). Mutations conferring heat-stable resistance could either lock the R protein in the active conformation or retain positive protein–protein interactions at high temperatures so that the R protein can be transported into or retained in the nucleus. This hypothesis is supported by the fact that each *R* gene can have a different temperature threshold for its activation of immunity. *R* genes *SNC1*, *RPS4*, and *N* are active at 22°C but inactive at 28°C. *R* genes responsible for hybrid necrosis and the auto-active form of the *RPP4 R* gene are active at 16°C but inactive at 22°C. The ultimate test of this conformation/interaction model will come from solving the structure of R proteins and their protein complexes at different temperatures.

There appear to be differences in the extent of temperature modulation of disease resistance, depending on the type of NB-LRR R proteins involved. Resistance mediated by the TIR type of R proteins is often more sensitive to heat than that of the CC type (Wang et al. 2009). This could result from structural differences or from altered modes of signaling between R protein subtypes. The TIR proteins require EDS1/PAD4 and usually need nuclear localization for their full function, while CC proteins require NDR1 rather than EDS1/PAD4 and nuclear localization is not always required. Further elucidation of the mechanisms of activation and signaling in R protein subgroups should shed light on their differential temperature sensitivity.

7.4.2 Involvement of NB-LRR R proteins in heat-sensitive immune responses

Temperature has been found to modulate some key signaling components in plant immunity. For instance, *EDS1* and *PAD4*, two regulators of both basal and *R*-mediated disease resistance, have higher steady RNA expression at 22°C than at 28°C (Yang and Hua 2004). Furthermore, SA accumulates to greater levels at low temperature (Scott et al. 2004). Given the observed temperature sensitivity of R proteins, it is possible that low-temperature induction of signaling molecules is a consequence of R activity rather than the basis of temperature-sensitive immunity. Indeed, upregulating *EDS1* and *PAD4* at the transcript level is not sufficient to induce resistance at high temperature (Wang et al. 2009). Nevertheless, these changes of core signaling components will reinforce *R* gene-induced temperature sensitivity and contribute to a general enhancement of SA-mediated resistance at low temperature.

Temperature sensitivity of NB-LRR R proteins may also account for heat sensitivity in many disease resistance systems not originally associated with NB-LRR proteins. Cf-4- and C-f9-mediated resistance to *C. fulvum* is suppressed at a high temperature (de Jong et al. 2002). Although an as-yet unidentified component might control this temperature sensitivity, a recent study has shown that Cf-4 and Cf-9 both require the NB-LRR protein AVR-RESPONSIVE TOMATO (ART) for HR and disease resistance (Gabriels et al. 2007). The ART protein could therefore potentially be responsible for temperature sensitivity in Cf-mediated resistance. In a similar case, the NB-LRR protein Prf is required for the cytoplasmic kinase R protein Pto to induce immune responses against bacterial pathogens with effectors AvrPto or AvrPtoB (Salmeron et al. 1996; Mucyn et al. 2006). NB-LRR proteins can function in pairs to mediate disease resistance (Eitas and Dangl 2010). These can include both TIR and CC types, such as the N and NRG1 pair and the RPM1 and TAO1 pair (Peart et al. 2005; Eitas et al. 2008). The role of NB-LRR proteins therefore extends beyond pathogen perception, suggesting a more prevalent involvement of these proteins in immunity.

The heat sensitivity of *R* gene activation and signaling is also likely to account for the temperature-dependent phenotypes of a number of autoimmune mutants. For example, *SNC1* is responsible for heat sensitivity of the autoimmune phenotype of the *bon1* mutant (Hua et al. 2001; Yang and Hua 2004). A large number of lesion mimic or autoimmune mutants have not yet been cloned, but their heat sensitivity suggests the involvement of *R* genes.

7.4.3 Enhancement of immunity by ABA deficiency at high temperatures

The inhibition of *R*-mediated resistance at high temperature is antagonized by deficiency in the plant hormone abscisic acid (ABA). In the mutant screen for heat-stable disease resistance discussed earlier, two genes, *ABA2* and *ABA3*, were identified (Mang et al. 2012; Zhu et al. 2012). The ABA-deficient mutants, *aba2* and *aba3*, displayed SNC1-mediated resistance at 28°C, conditions where resistance is normally suppressed. Interestingly, ABA deficiency promotes the nuclear accumulation of SNC1, while ABA addition acts oppositely. The effect of ABA on SNC1 localization is independent on SA which is often thought to be antagonized by ABA. ABA deficiency has also been shown to enhance the activity and nuclear localization of RPS4 at 28°C, suggesting a broad regulation of R proteins with a nuclear activity. Because ABA concentration is not affected by temperature, it may function independently to high-temperature signaling to inhibit the nuclear accumulation and activity of R proteins (Figure 7.1). How this process occurs has yet to be discovered.

7.4.4 Cold sensitivity in RNA silencing-mediated immunity

Antiviral RNA silencing can be induced by high temperature or inhibited by low temperature. RDR6 has been shown to mediate antiviral silencing to three RNA viruses in a response stimulated by high growth temperature (Qu et al. 2005). DCL2, AGO2, and HEN1 are critical for silencing of turnip crinkle virus (TCV), especially at high temperatures. Furthermore, the *dcl2*, *ago2*, and *hen1* mutants have a more pronounced susceptibility to viruses at 26°C than at 18°C (Zhang et al. 2012a). It has been suggested that RDR6 and DCL2 may be activated at high temperature due to a greater accumulation of their product siRNAs in these conditions (Qu et al. 2005; Zhang et al. 2012a). This hypothesis awaits validation from further biochemical and genetic studies. Surprisingly, RNA silencing of endogenous plant genes using geminivirus-mediated virus-induced gene silencing (VIGS) was found to be enhanced rather than suppressed by low temperature (Tuttle et al.

2008). Further study should reveal if contrasting effects on viral silencing and endogenous gene silencing arise from differences in biogenesis of siRNAs or from effects of viral silencing suppressors.

7.5 Evolution of the temperature sensitivity of immunity

Why are plant immune response often effective at one temperature but not another? Is immunity an evolving process, or is temperature a biophysical or biochemical constraint, restricting immunity to a certain temperature range? Although there are no definite answers to these questions, high fitness costs may have placed immunity under very tight controls. A hypothesis on the evolution of temperature sensitivity in plant immunity is presented in Figure 7.2. This requires greater understanding of plant and pathogen systems to validate, refine, or even disapprove. In this model, plants prioritize growth, abiotic stress responses, and biotic stress responses based on environmental cues. Biotic stress positively parallels with plant immunity, while abiotic stress behaves oppositely. Temperature effects on

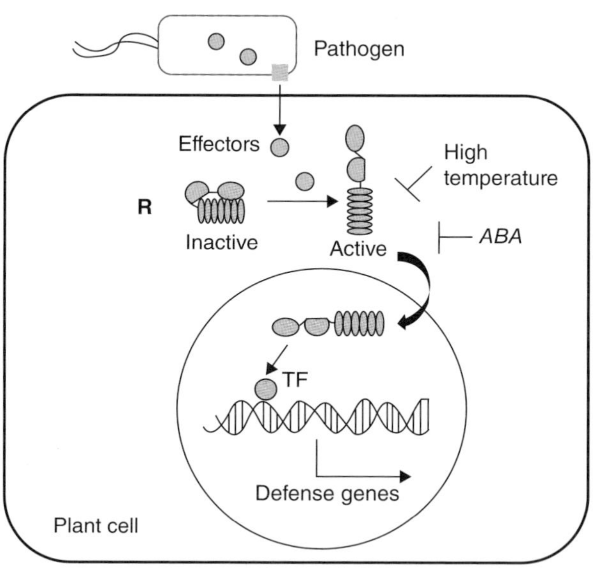

Figure 7.2 Hypothesis on the evolution of temperature modulation of plant immunity. Temperature affects pathogen virulence and abiotic responses in plants. Plant immunity is positively correlated with biotic stresses and inversely correlated with abiotic stresses. Temperature sensitivity of plant immunity is likely shaped by the integrated information of biotic and abiotic pressures inferred from temperature signals.

these stresses may have shaped the sensitivity of immunity to temperature (Figure 7.2).

7.5.1 Coevolution with pathogens

Plant pathogenic bacteria, unlike their human counterparts, can be more virulent at lower temperatures. Some virulence genes exhibit increased transcription at temperatures much lower than their optimal growth temperature. Virulence determinants include phytotoxins, cell-wall degrading enzymes, and the bacterial type III secretion system (Smirnova et al. 2001). The production of coronatine by virulent *P. syringae* is maximal at 18°C and minimal at 28°C (Palmer and Bender 1993; Weingart et al. 2004). Strikingly, several bacterial effector proteins are secreted most abundantly at temperatures between 18°C and 22°C (van Dijk et al. 1999). Weaker virulence at high temperatures could therefore be a strategy for bacteria to replicate at optimal growth temperatures. This would then exert less pressure for plants to maintain *R*-mediated resistance at high temperatures.

Exceptions to high-temperature inhibition of *R* resistance exist and may have arisen from pathogen pressures. Resistance against root-knot nematode mediated by *Mi* genes is usually suppressed by temperatures above 28°C. The *Mi-9* gene, however, conferred resistance at 28°C. Although not yet cloned, *Mi-9* is in a gene cluster containing multiple *R* genes homologous to *Mi-1* which confers a heat-sensitive resistance (Jablonska et al. 2007). It is likely that the *Mi-9* gene has evolved from a heat-sensitive *R* gene by a few nucleotide changes, suggesting that heat-stable resistance could arise in nature when pathogen pressure is high. Another similar scenario is resistance against rice bacterial blight conferred by different *Xa* genes. Although most are more effective in cooler conditions, *Xa7*-conferred resistance is more efficient at high temperature (Webb et al. 2010). Most *Xa* genes are not yet cloned, but it is possible *Xa7* could have been evolved from a heat-sensitive *R* gene due to selective pressures from pathogens at high temperature.

The induction of antiviral RNA silencing at high temperatures may have originally resulted from an enhanced replication rate of viruses in these conditions. It has been reported that TCV produces more viral RNAs at 26°C than at 18°C in the presence of viral silencing (Zhang et al. 2012a). It is not yet known if this is a widespread feature as viral replication has also been shown to be reduced at high temperature. This discrepancy may have arisen from the fact that the extent of viral replication is the outcome of virulence and plant immune responses combined. In this way, a low viral replication may indicate an effective antiviral response rather than a low virulence. Alternatively, a reduction of *R*-mediated resistance at high temperatures could have resulted in plants evolving enhanced RNA silencing in these conditions.

7.5.2 Competition between biotic and abiotic responses

Temperature sensitivity of plant immunity may reflect a complex interaction between biotic and abiotic responses. Both HSP90 and HSC70 are involved in plant defense responses by performing chaperone functions together with *SUPPRESSOR OF THE G2 ALLELE OF SKP1* (SGT1) and *REQUIRED FOR MLA12 RESISTANCE 1* (RAR1) (Noel et al. 2007; Kadota et al. 2010). These chaperone proteins could be recruited to perform more essential cellular functions at elevated temperatures, thereby compromising plant defense responses. Overexpression of HSC70 has been shown to enhance heat tolerance and disable both basal and *R*-mediated disease resistance (Noel et al. 2007). These observations support the hypothesis of a trade-off between biotic and abiotic responses.

Reports that ABA deficiency can promote immunity at high temperature also suggest competition between abiotic and biotic responses. ABA has diverse roles in plants and its level often indicates the level of abiotic stress (Cutler et al. 2010). A low ABA level may allow plants to shift the balance from abiotic response toward biotic responses, thus potentiating immunity at high temperature.

7.6 Concluding remarks

In summary, temperature has a profound effect on plant immunity. Plants use environmental cues to infer and anticipate pathogen and abiotic stress pressures to effectively balance growth and defense. The significant effect of temperature on plant immune systems may have facilitated the diversification of key immune components such as R proteins by masking their potentially deleterious effects during the evolutionary process. As autoimmunity could have acted as a gene-flow barrier in plant species (Bomblies and Weigel 2007), temperature sensitivity may have shaped the evolution of speciation by facilitating *R* gene diversification. With greater understanding of the temperature sensitivity of plant immunity at the molecular level, we will be in a better position to predict and cope with effects of climate change on plant disease and disease resistance.

References

Alcazar R, Garcia AV, Parker JE, Reymond M (2009) Incremental steps toward incompatibility revealed by *Arabidopsis* epistatic interactions modulating salicylic acid pathway activation. Proc. Natl. Acad. Sci. USA 106: 334–339.

Alvarado V, Scholthof HB (2009) Plant responses against invasive nucleic acids: RNA silencing and its suppression by plant viral pathogens. Semin. Cell Dev. Biol. 20: 1032–1040.

Ausubel FM (2005) Are innate immune signaling pathways in plants and animals conserved? Nat. Immunol. 6: 973–979.

Baulcombe D (2002) Viral suppression of systemic silencing. Trends Microbiol. 10: 306–308.

Bhattacharjee S, Halane MK, Kim SH, Gassmann W (2011) Pathogen effectors target *Arabidopsis* EDS1 and alter its interactions with immune regulators. Science 334: 1405–1408.

Boller T, Felix G (2009) A renaissance of elicitors: Perception of microbe-associated molecular patterns and danger signals by pattern-recognition receptors. Annu. Rev. Plant Biol. 60: 379–406.

Bomblies K, Weigel D (2007) Hybrid necrosis: Autoimmunity as a potential gene-flow barrier in plant species. Nat. Rev. Genet. 8: 382–393.

Bomblies K, Lempe J, Epple P, Warthmann N, Lanz C, Dangl JL, Weigel D (2007) Autoimmune response as a mechanism for a Dobzhansky–Muller-type incompatibility syndrome in plants. PLoS Biol. 5: e236.

Burch-Smith TM, Schiff M, Caplan JL, Tsao J, Czymmek K, Dinesh-Kumar SP (2007) A novel role for the TIR domain in association with pathogen-derived elicitors. PLoS Biol. 5: e68.

Caplan J, Padmanabhan M, Dinesh-Kumar SP (2008) Plant NB-LRR immune receptors: From recognition to transcriptional reprogramming. Cell Host Microbe. 3: 126–135.

Chellappan P, Vanitharani R, Ogbe F, Fauquet CM (2005) Effect of temperature on geminivirus-induced RNA silencing in plants. Plant Physiol. 138: 1828–1841.

Chisholm ST, Coaker G, Day B, Staskawicz BJ (2006) Host–microbe interactions: Shaping the evolution of the plant immune response. Cell 124: 803–814.

Cutler SR, Rodriguez PL, Finkelstein RR, Abrams SR (2010) Abscisic acid: Emergence of a core signaling network. Annu. Rev. Plant Biol. 61: 651–679.

Dangl JL, Jones JD (2001) Plant pathogens and integrated defence responses to infection. Nature 411: 826–833.

Deslandes L, Rivas S (2012) Catch me if you can: Bacterial effectors and plant targets. Trends Plant Sci. 17: 644–655.

Dhaka A, Viswanath V, Patapoutian A (2006) Trp ion channels and temperature sensation. Annu. Rev. Neurosci. 29: 135–161.

van Dijk K, Fouts DE, Rehm AH, Hill AR, Collmer A, Alfano JR (1999) The Avr (effector) proteins HrmA (HopPsyA) and AvrPto are secreted in culture from *Pseudomonas syringae* pathovars via the Hrp (type III) protein secretion system in a temperature- and pH-sensitive manner. J. Bacteriol. 181: 4790–4797.

Ding SW (2010) RNA-based antiviral immunity. Nat. Rev. Immunol. 10: 632–644.

Dodds PN, Rathjen JP (2010) Plant immunity: Towards an integrated view of plant–pathogen interactions. Nat. Rev. Genet. 11: 539–548.

Dropkin V (1969) The necrotic reaction of tomatoes and other hosts resistant to Meloidogyne: Reversal by temperature. Phytopathology 59: 1632–1637.

Durrant WE, Dong X (2004) Systemic acquired resistance. Annu. Rev. Phytopathol. 42: 185–209.

Eitas TK, Dangl JL (2010) NB-LRR proteins: Pairs, pieces, perception, partners, and pathways. Curr. Opin. Plant Biol. 13: 472–477.

Eitas TK, Nimchuk ZL, Dangl JL (2008) *Arabidopsis* TAO1 is a TIR-NB-LRR protein that contributes to disease resistance induced by the *Pseudomonas syringae* effector AvrB. Proc. Natl. Acad. Sci. USA 105: 6475–6480.

Elmore JM, Lin ZJ, Coaker G (2011) Plant NB-LRR signaling: Upstreams and downstreams. Curr. Opin. Plant Biol. 14: 365–371.

Flor HH (1971) Current status of the gene-for-gene concept. Annu. Rev. Phytopathol. 9: 275–296.

Fu D, Uauy C, Distelfeld A, Blechl A, Epstein L, Chen X, Sela H, Fahima T, Dubcovsky J (2009) A kinase-START gene confers temperature-dependent resistance to wheat stripe rust. Science 323: 1357–1360.

Gabriels SHEJ, Vossen JH, Ekengren SK, van Ooijen G, Abd-El-Haliem AM, van den Berg GCM, Rainey DY, Martin GB, Takken FLW, de Wit PJGM, Joosten MHAJ (2007) An NB-LRR protein required for HR signalling mediated by both extra- and intracellular resistance proteins. Plant J. 50: 14–28.

Garrett KA, Dendy SP, Frank EE, Rouse MN, Travers SE (2006) Climate change effects on plant disease: Genomes to ecosystems. Annu. Rev. Phytopathol. 44: 489–509.

Gaudet DA, Tronsmo A, Laroche A (2011) Climate change and plant diseases. In: Storey KB, Tanino KK (eds). Temperature adaptation in a changing climate: Nature at risk. CABI, Wallingford, pp. 144–159.

Glazebrook J (2001) Genes controlling expression of defense responses in *Arabidopsis*: 2001 status. Curr. Opin. Plant Biol. 4: 301–308.

Glazebrook J (2005) Contrasting mechanisms of defense against biotrophic and necrotrophic pathogens. Annu. Rev. Phytopathol. 43: 205–227.

Greenberg JT, Yao N (2004) The role and regulation of programmed cell death in plant–pathogen interactions. Cell. Microbiol. 6: 201–211.

Hammond-Kosack KE, Jones JD (1996) Resistance gene-dependent plant defense responses. Plant Cell 8: 1773–1791.

Heidrich K, Wirthmueller L, Tasset C, Pouzet C, Deslandes L, Parker JE (2011) *Arabidopsis* EDS1 connects pathogen effector recognition to cell compartment-specific immune responses. Science 334: 1401–1404.

Heidrich K, Blanvillain-Baufume S, Parker JE (2012) Molecular and spatial constraints on NB-LRR receptor signaling. Curr. Opin. Plant Biol. 15: 385–391.

Hogenhout SA, Bos JIB (2011) Effector proteins that modulate plant–insect interactions. Curr. Opin. Plant Biol. 14: 422–428.

Howe GA, Jander G (2008) Plant immunity to insect herbivores. Annu. Rev. Plant Biol. 59: 41–66.

Hua J, Grisafi P, Cheng SH, Fink GR (2001) Plant growth homeostasis is controlled by the *Arabidopsis BON1* and *BAP1* genes. Genes Dev. 15: 2263–2272.

Huang X, Li J, Bao F, Zhang X, Yang S (2010) A gain-of-function mutation in the *Arabidopsis* disease resistance gene RPP4 confers sensitivity to low temperature. Plant Physiol. 154: 796–809.

Hwang CF, Bhakta AV, Truesdell GM, Pudlo WM, Williamson VM (2000) Evidence for a role of the N terminus and leucine-rich repeat region of the Mi gene product in regulation of localized cell death. Plant Cell. 12: 1319–1329.

Ichimura K, Casais C, Peck SC, Shinozaki K, Shirasu K (2006) MEKK1 is required for MPK4 activation and regulates tissue-specific and temperature-dependent cell death in *Arabidopsis*. J. Biol. Chem. 281: 36969–36976.

Jablonska B, Ammiraju JS, Bhattarai KK, Mantelin S, Martinez de Ilarduya O, Roberts PA, Kaloshian I (2007) The Mi-9 gene from *Solanum arcanum* conferring heat-stable resistance to root-knot nematodes is a homolog of Mi-1. Plant Physiol. 143: 1044–1054.

Johansson J, Mandin P, Renzoni A, Chiaruttini C, Springer M, Cossart P (2002) An RNA thermosensor controls expression of virulence genes in *Listeria monocytogenes*. Cell 110: 551–561.

Jones JD, Dangl JL (2006) The plant immune system. Nature 444: 323–329.

de Jong CF, Takken FL, Cai X, de Wit PJ, Joosten MH (2002) Attenuation of Cf-mediated defense responses at elevated temperatures correlates with a decrease in elicitor-binding sites. Mol. Plant Microbe Interact. 15: 1040–1049.

Kadota Y, Shirasu K, Guerois R (2010) NLR sensors meet at the SGT1-HSP90 crossroad. Trends Biochem. Sci. 35: 199–207.

Kim SH, Gao F, Bhattacharjee S, Adiasor JA, Nam JC, Gassmann W (2010) The *Arabidopsis* resistance-like gene SNC1 is activated by mutations in SRFR1 and contributes to resistance to the bacterial effector AvrRps4. PLoS Pathog. 6: e1001172.

Lukasik E, Takken FL (2009) STANDing strong, resistance proteins instigators of plant defence. Curr. Opin. Plant Biol. 12: 427–436.

Mang HG, Qian W, Zhu Y, Qian J, Kang HG, Klessig D, Hua J (2012) Abscisic acid deficiency antagonizes high-temperature inhibition of disease resistance through enhancing nuclear accumulation of resistance proteins SNC1 and RPS4 in *Arabidopsis*. Plant Cell 24: 1271–1284.

Martin G, Bogdanove A, Sessa G (2003) Understanding the functions of plant disease resistance proteins. Annu. Rev. Plant Biol. 54: 23–61.

Meyers BC, Kozik A, Griego A, Kuang H, Michelmore RW (2003) Genome-wide analysis of NBS-LRR-encoding genes in *Arabidopsis*. Plant Cell 15: 809–834.

Monaghan J, Zipfel C (2012) Plant pattern recognition receptor complexes at the plasma membrane. Curr. Opin. Plant Biol. 15: 349–357.

Monaghan J, Germain H, Weihmann T, Li X (2010) Dissecting plant defence signal transduction: Modifiers of snc1 in *Arabidopsis*. Can. J. Plant Pathol. 32: 35–42.

Moyer MM, Gadoury DM, Cadle-Davidson L, Dry IB, Magarey PA, Wilcox WF, Seem RC (2010) Effects of acute low-temperature events on development of *Erysiphe necator* and susceptibility of *Vitis vinifera*. Phytopathology 100: 1240–1249.

Mucyn TS, Clemente A, Andriotis VM, Balmuth AL, Oldroyd GE, Staskawicz BJ, Rathjen JP (2006) The tomato NBARC-LRR protein Prf interacts with Pto kinase in vivo to regulate specific plant immunity. Plant Cell 18: 2792–2806.

Noel LD, Cagna G, Stuttmann J, Wirthmuller L, Betsuyaku S, Witte CP, Bhat R, Pochon N, Colby T, Parker JE (2007) Interaction between SGT1 and cytosolic/nuclear HSC70 chaperones regulates *Arabidopsis* immune responses. Plant Cell 19: 4061–4076.

Nurnberger T, Brunner F, Kemmerling B, Piater L (2004) Innate immunity in plants and animals: Striking similarities and obvious differences. Immunol. Rev. 198: 249–266.

Oliver RP, Solomon PS (2010) New developments in pathogenicity and virulence of necrotrophs. Curr. Opin. Plant Biol. 13: 415–419.

Palmer DA, Bender CL (1993) Effects of environmental and nutritional factors on production of the polyketide phytotoxin coronatine by *Pseudomonas syringae* pv. glycinea. Appl. Environ. Microbiol. 59: 1619–1626.

Peart JR, Mestre P, Lu R, Malcuit I, Baulcombe DC (2005) NRG1, a CC-NB-LRR protein, together with N, a TIR-NB-LRR protein, mediates resistance against tobacco mosaic virus. Curr. Biol. 15: 968–973.

Qu F, Ye XH, Hou GC, Sato S, Clemente TE, Morris TJ (2005) RDR6 has a broad-spectrum but temperature-dependent antiviral defense role in *Nicotiana benthamiana*. J. Virol. 79: 15209–15217.

Rafiqi M, Bernoux M, Ellis JG, Dodds PN (2009) In the trenches of plant pathogen recognition: Role of NB-LRR proteins. Semin. Cell Dev. Biol. 20: 1017–1024.

Rafiqi M, Ellis JG, Ludowici VA, Hardham AR, Dodds PN (2012) Challenges and progress towards understanding the role of effectors in plant–fungal interactions. Curr. Opin. Plant Biol. 15: 477–482.

Salmeron JM, Oldroyd GED, Rommens CMT, Scofield SR, Kim HS, Lavelle DT, Dahlbeck D, Staskawicz BJ (1996) Tomato Prf is a member of the leucine-rich repeat class of plant disease resistance genes and lies embedded within the Pto kinase gene cluster. Cell 86: 123–133.

Samuel G (1931) Some experiments on inoculating methods with plant viruses, and on local lesions. Ann. Appl. Biol. 18: 494–507.

Scheel D (1998) Resistance response physiology and signal transduction. Curr. Opin. Plant Biol. 1: 305–310.

Scott IM, Clarke SM, Wood JE, Mur LA (2004) Salicylate accumulation inhibits growth at chilling temperature in *Arabidopsis*. Plant Physiol. 135: 1040–1049.

Shen QH, Saijo Y, Mauch S, Biskup C, Bieri S, Keller B, Seki H, Ulker B, Somssich IE, Schulze-Lefert P (2007) Nuclear activity of MLA immune receptors links isolate-specific and basal disease-resistance responses. Science 315: 1098–1103.

Silhavy D, Burgyan J (2004) Effects and side-effects of viral RNA silencing suppressors on short RNAs. Trends Plant Sci. 9: 76–83.

Smirnova A, Li H, Weingart H, Aufhammer S, Burse A, Finis K, Schenk A, Ullrich MS (2001) Thermoregulated expression of virulence factors in plant-associated bacteria. Arch. Microbiol. 176: 393–399.

Soosaar JL, Burch-Smith TM, Dinesh-Kumar SP (2005) Mechanisms of plant resistance to viruses. Nat. Rev. Microbiol. 3: 789–798.

Spoel SH, Johnson JS, Dong X (2007) Regulation of tradeoffs between plant defenses against pathogens with different lifestyles. Proc. Natl. Acad. Sci. USA 104: 18842–18847.

Szittya G, Silhavy D, Molnar A, Havelda Z, Lovas A, Lakatos L, Banfalvi Z, Burgyan J (2003) Low temperature inhibits RNA silencing-mediated defence by the control of siRNA generation. EMBO J. 22: 633–640.

Tuttle JR, Idris AM, Brown JK, Haigler CH, Robertson D (2008) Geminivirus-mediated gene silencing from cotton leaf crumple virus is enhanced by low temperature in cotton. Plant Physiol. 148: 41–50.

Vlot AC, Dempsey DA, Klessig DF (2009) Salicylic acid, a multifaceted hormone to combat disease. Annu. Rev. Phytopathol. 47: 177–206.

Wang Y, Bao Z, Zhu Y, Hua J (2009) Analysis of temperature modulation of plant defense against biotrophic microbes. Mol. Plant Microbe Interact. 22: 498–506.

Wang MB, Masuta C, Smith NA, Shimura H (2012) RNA silencing and plant viral diseases. Mol. Plant Microbe Interact. 25: 1275–1285.

Webb KM, Ona I, Bai J, Garrett KA, Mew T, Vera Cruz CM, Leach JE (2010) A benefit of high temperature: Increased effectiveness of a rice bacterial blight disease resistance gene. New Phytol. 185: 568–576.

Weingart H, Stubner S, Schenk A, Ullrich MS (2004) Impact of temperature on in planta expression of genes involved in synthesis of the *Pseudomonas syringae* phytotoxin coronatine. Mol. Plant Microbe Interact. 17: 1095–1102.

Wiermer M, Feys BJ, Parker JE (2005) Plant immunity, the EDS1 regulatory node. Curr. Opin. Plant Biol. 8: 383–389.

Whitham S, McCormick S, Baker B (1996) The N gene of tobacco confers resistance to tobacco mosaic virus in transgenic tomato. Proc. Natl. Acad. Sci. USA 93: 8776–8781.

Wirthmueller L, Zhang Y, Jones JD, Parker JE (2007) Nuclear accumulation of the *Arabidopsis* immune receptor RPS4 is necessary for triggering EDS1-dependent defense. Curr. Biol. 17: 2023–2029.

Xiao S, Brown S, Patrick E, Brearley C, Turner JG (2003) Enhanced transcription of the *Arabidopsis* disease resistance genes *RPW8.1* and *RPW8.2* via a salicylic acid-dependent amplification circuit is required for hypersensitive cell death. Plant Cell 15: 33–45.

Yang S, Hua J (2004) A haplotype-specific resistance gene regulated by *BONZAI1* mediates temperature-dependent growth control in *Arabidopsis*. Plant Cell 16: 1060–1071.

Yang H, Shi Y, Liu J, Guo L, Zhang X, Yang S (2010) A mutant CHS3 protein with TIR-NB-LRR-LIM domains modulates growth, cell death and freezing tolerance in a temperature-dependent manner in *Arabidopsis*. Plant J. 63: 283–296.

Zhang Y, Goritschnig S, Dong X, Li X (2003) A gain-of-function mutation in a plant disease resistance gene leads to constitutive activation of downstream signal transduction pathways in *suppressor of npr1-1, constitutive 1*. Plant Cell 15: 2636–2646.

Zhang X, Singh J, Li D, Qu F (2012a) Temperature-dependent survival of Turnip crinkle virus-infected *Arabidopsis* plants relies on an RNA silencing-based defense that requires dcl2, AGO2, and HEN1. J. Virol. 86: 6847–6854.

Zhang Z, Wu Y, Gao M, Zhang J, Kong Q, Liu Y, Ba H, Zhou J, Zhang Y (2012b) Disruption of PAMP-induced MAP kinase cascade by a *Pseudomonas syringae* effector activates plant immunity mediated by the NB-LRR protein SUMM2. Cell Host Microbe. 11: 253–263.

Zhu Y, Qian W, Hua J (2010a) Temperature modulates plant defense responses through NB-LRR proteins. PLoS Pathog. 6: e1000844.

Zhu Z, Xu F, Zhang Y, Cheng YT, Wiermer M, Li X (2010b) *Arabidopsis* resistance protein SNC1 activates immune responses through association with a transcriptional corepressor. Proc. Natl. Acad. Sci. USA 107: 13960–13965.

Zhu Y, Mang HG, Sun Q, Qian J, Hipps A, Hua J (2012) Gene discovery using mutagen-induced polymorphisms and deep sequencing: Application to plant disease resistance. Genetics 192: 139–146.

8 Temperature, climate change, and global food security

Robert J. Redden, Jerry L. Hatfield, P.V. Vara Prasad, Andreas W. Ebert, Shyam S. Yadav, and Garry J. O'Leary

8.1 Introduction

The world is undergoing changes unprecedented since the first transition from a hunter-gatherer population to agriculture 11,000 years ago. Technologies such as irrigation, metal working, urban development, and medicine developed slowly at first, then rapidly expanded with the industrial revolution after the eighteenth century AD, with increased human longevity, a population increasing beyond seven billion, and vastly increased consumption of fossil fuels. The cost, however, of this extraordinary progress is multifaceted pollution, particularly of increased levels of atmospheric CO_2, methane, sulfur, and aerosols, which are having very complex interactions with the global atmosphere and ultimately on climate (IPCC 2007a). The net result is a continuing increase in global temperature, which is unlikely to be mitigated in the short term as energy conservation measures are overwhelmed by a large increase in the human footprint of pollution with around a 30% increase in population predicted by 2050.

Not only will temperature increase globally by 0.8–1.0°C by 2050 and further increase by 2100, but temperature variability will increase with more severe high-temperature spikes during crop growing seasons than previously experienced. These effects will severely affect agricultural production systems and hence food security (Lobell et al. 2011), the foundation upon which modern society depends. This chapter examines the effect of increasing temperature on plant growth and cropping systems, the implications for food production of various crops from a global perspective, the challenges for human nutrition, and the opportunities to modify and adapt cropping systems to climate change.

8.2 Climate change on a global basis

Temperature and precipitation are two critical parameters affecting plant development. Changes in these factors may induce plant stress through direct effects on plant growth, increased water use resulting from increased

Temperature and Plant Development, First Edition. Edited by Keara A. Franklin and Philip A. Wigge.

evaporative demand, or drought responses induced by altered precipitation patterns. In terms of temperature change, Meehl et al. (2007) state that on a global basis,

> it is very likely that heat waves will be more intense, more frequent, and longer lasting in a future warmer climate. Cold episodes are projected to decrease significantly in a future warmer climate. Almost everywhere, daily minimum temperatures are projected to increase faster than daily maximum temperatures, leading to a decrease in diurnal temperature range. Decreases in frost days are projected to occur almost everywhere in the middle and high latitudes, with a comparable increase in growing season length (10.ES).

It is expected that by mid-century, 2040–2050, air temperatures will increase on average around the globe between 0.8°C and 1.0°C and by the end of the century have increased between 2°C and 4°C. During this period, the emerging trend is for a proportionally larger increase in average minimum temperatures than in average maximum temperatures with the greater chance for heat stress conditions occurring during the summer months. These trends have already begun to exhibit themselves in the climate record around the globe as reported by Lobell et al. (2011). Such temperature changes will have a significant impact on crop growth and development.

In their assessment on future trends in precipitation, Meehl et al. (2007) stated:

> For a future warmer climate, the current generation of models indicates that precipitation generally increases in the areas of regional tropical precipitation maxima (such as the monsoon regimes) and over the tropical Pacific in particular, with general decreases in the subtropics, and increases at high latitudes as a consequence of a general intensification of the global hydrological cycle. Globally averaged mean water vapour, evaporation, and precipitation are projected to increase (10.ES).

Precipitation changes are critical to agriculture, and shifts in both amounts and timing could have serious implications for agricultural production, disrupting all cultural operations from planting to harvesting.

Changes in climate are associated with changing levels of greenhouse gases in the atmosphere including water vapor. Katz and Brown (1992) report that the increased variability in climate attributes may be more important than increases in mean values. Continuing efforts to update projected climate scenarios aim to facilitate the assessment of potential future impacts for the IPCC Fifth Assessment Report. Projections of climate change are often met with skepticism because of the natural variation that occurs in seasonal weather between successive years. There are, however, definite signals that climate is indeed changing (Karl et al. 2009). There is an unmistakable increase in carbon dioxide (CO_2) concentrations in the atmosphere along with nitrous oxide (N_2O) and methane (CH_4) (IPCC 2007a). These levels have now reached concentrations exceeding all recorded values in the history of cultivated agriculture.

The causes of climate change can be linked to the increased impact of human activities on the concentration of greenhouse gases, increases in aerosol content in the atmosphere, and changes in land use patterns. These effects influence the radiation balance of the Earth, evaporation rate from the Earth's surface, and patterns of heating and cooling around the globe. Several analyses have described the impact of human activity on climate change (IPCC 2007b). The negative effects of climate change on agriculture will be most pronounced in developing countries (IPCC 2007b; Nelson et al. 2009, 2010; Wassmann et al. 2010; Lobell et al. 2011; Müller et al. 2011; de la Peña et al. 2011). A poleward shift of crop zones due to increased average temperatures might, however, have favorable effects on agriculture in the northernmost regions of continental Asia (FAO 2004) and possibly Eastern Europe. Projections for the near term (20–30 years) predict that climate variability and extreme weather events will increase and affect all regions with increasing negative impacts on growth and yield, leading to increased concerns about food security, particularly in sub-Saharan Africa and South Asia (Burney et al. 2010; IPCC 2012).

8.3 The impact of temperature on crop water relations

Plant responses to climate occur at both the leaf and canopy level and are expressed in responses ranging from photosynthetic and transpiration rates to leaf expansion. Temperature and precipitation are the two primary factors affecting plant growth, with temperature effects separated into two aspects, direct effects on plant growth responses and indirect effects on plant water use rates mediated by changes in atmospheric demand.

The maximum amount of water that can be held as vapor in the atmosphere is termed saturation vapor pressure which is a direct exponential function of air temperature (Buck 1981). As temperature increases, the amount of water that can be held as vapor will therefore increase. The difference between the actual amount of water in the atmosphere (actual vapor pressure) and the saturated vapor pressure is termed the vapor pressure deficit, and this determines the evaporative demand on the crop and resultant water use. As temperature increases, potential water use will increase exponentially. The likelihood for more variable precipitation may therefore lead to increased water stress in plants during the growing season. Leaf and canopy level feedbacks due to increased air temperature and atmospheric demand have been detailed in Hatfield and Prueger (2011). This study suggests that both leaf and canopy level responses are a complex set of interactions between the microclimatic variables, for example, temperature, atmospheric water content, solar radiation, and wind speed, and the canopy/leaf parameters, for example, leaf shape, leaf size, stomatal conductance, leaf orientation, canopy size, and canopy leaf arrangement. A canopy with a small number of leaves will exhibit a greater response to changing atmospheric demand because each leaf is more closely coupled with the atmosphere. A denser canopy with more leaves will, conversely, show less tightly coupled responses with the atmospheric environment. The larger and denser canopy will begin to create its own microclimate surrounding the leaves and show less responsiveness to the atmospheric changes. It is therefore possible that a dense canopy can act as mulch, reducing evaporation from the soil surface (O'Leary et al. 1985). This coupling is the primary reason that canopy water use increases rapidly with canopy size and then begins to plateau when the leaf area index increases above 4. A similar response is found for light interception – which rapidly increases as the canopy develops and then reaches a plateau when leaf area index approaches 4 and above.

An often overlooked aspect in evaluating plant response to climate is the fact that air temperature doesn't simply equate to leaf temperature or canopy temperature (T_c), so the actual temperature that the plant experiences may be different from the air temperature. The degree of difference between leaf and air temperature was first observed by Tanner (1963) and further expanded by Wiegand and Namken (1966) to produce a relationship between leaf temperature and plant water status. Leaf temperature or T_c of well-watered canopies is often 3–5°C less than air temperature because of evaporative cooling from transpiration through the leaf surface. As available soil water decreases, leaf temperature more closely tracks air temperature and under extreme water stress may exceed air temperature. Approaches to quantify crop water stress using T_c are now well established (Hatfield et al. 2008).

An interesting development in our understanding of plant responses to the environment has been the linkage between temperature responses of leaves and the thermal stability of metabolic enzymes (Burke et al. 1988). This work suggests that plant leaves have an optimum temperature for metabolic performance based on the metabolic activity of enzymes. A recent study reported by Conaty et al. (2012) observed that cotton (*Gossypium hirsutum* L.) carbon assimilation, measured with leaf photosynthetic chambers, peaked at a T_c of 29°C, while the fluorescence assay was optimal between 28°C and 30°C, and yield was maximum at 26°C. When T_c exceeded 28°C there was a decrease in yield. Linkages between air temperature, T_c, and crop water status have been described in Hatfield and Prueger (2011) and suggest that under well-watered conditions, plants can maintain leaf temperatures near their optimal temperature range. As soil water becomes limiting, however, leaf temperatures are no longer maintained within a plant's optimal range and stress will result. The complex dynamic between plant temperature, air temperature, and soil water availability is an underlying reason why changing maximum air temperatures have less of an impact on plant growth and development compared to minimum air temperatures. During the day, increasing maximum temperatures would be moderated by increasing crop water use, leading to an actual leaf temperature which is more closely aligned with the optimal plant temperature. At night, however, with no transpiration, leaf temperatures would be in equilibrium with the air temperature because the only energy exchange will be that of long-wave radiation between the canopy and the atmosphere. These relationships can be described in mathematical form (Hatfield and Prueger 2011). The management of crops at elevated temperature will therefore depend not only on the absolute climatic changes but also on the willingness of farmers to adapt their diverse farming systems to ones that conserve and utilize water more efficiently (O'Leary et al. 2011).

Many of the predictions about the effect of increasing temperature on crops have focused on average air temperature. Increases in minimum air temperature changes may, however, be of greater consequence to plant growth and phenology (Hatfield et al. 2011). Changes in minimum temperatures are more likely to occur over broad geographic scales when compared to maximum temperatures (Knowles et al. 2006). Maximum temperatures are affected more by local conditions, and the role of soil water content and evaporative heat loss as soil water evaporates is particularly important (Alfaro et al. 2006). The linkage between soil water, evaporation rate, and maximum air temperature demonstrates the role of land surface cover and adequate soil water evaporation on maximum air temperatures. Minimum temperatures would be affected by atmospheric water vapor content and cloud cover because these two factors directly affect long-wave radiation exchange. The respiration rate of plants is directly related to minimum air temperatures, with increased respiration reducing biomass accumulation and crop yield (Hatfield et al. 2011). The overall effect of changing maximum and minimum temperatures therefore needs to be examined more closely than changes in the average temperature. This is exemplified in a study by Welch et al. (2010) which showed that higher minimum temperatures in Asia reduced rice (*Oryza sativa* L.) yields, while higher maximum temperature raised yields. They further observed that maximum temperature seldom exceeded the critical optimum temperature for rice. Maximum temperatures were therefore predicted to decrease yields only if they rise substantially above a critical zone.

8.4 The influence of high temperature on crop physiology and yield processes

The impact of high temperature differs among crops and also genotypes within a crop species. The effect of high temperature on various physiological and yield processes depends on the intensity, rate of increase, and duration of the high-temperature exposure in addition to the stage of crop development. A simple measure of high-temperature stress on crop growth and development is not articulated well in the literature. Rather, the efficiencies of various processes appear to be affected in different ways. High temperature influences photosynthesis either by stomatal closure, thereby decreasing flow of CO_2 into mesophyll tissue, or by directly impairing the metabolic activities of key enzymes. Metabolic effects include decreases in ribulose 1,5-bisphosphate (RuBP) carboxylase/oxygenase (Rubisco) protein content, Rubisco activity, RuBP regeneration capacity, and Rubisco activase enzyme. High temperature also increases thylakoid membrane damage, resulting in decreased photosystem II (PSII) activity (Prasad et al. 2008a, b; Ristic et al. 2008; Prasad and Djanaguiraman 2011). Additional effects include dissociation of the oxygen-evolving complex, resulting in an imbalance in electron flow from the complex toward the acceptor side of the PSII reaction center. Respiration rates are generally increased at higher temperatures, although the regulation of respiration under high-temperature stress is less well understood.

Different species have their own unique temperature response curve (Hatfield et al. 2011). One general observation is that rising temperatures will accelerate the rate of development, ultimately resulting in smaller plants and reduced grain yield. Indeed, Lobell et al. (2011) project that grain yields have already began to decline due to exposure to temperatures above the optimum threshold. The expectation is for continued yield decrease for major crops resulting from increased temperature stress (Hatfield et al. 2011). In addition to plant growth, exposure of plants to temperatures above the upper threshold will also adversely affect plant pollination and reproductive processes (Klein et al. 2007; de la Peña et al. 2011; Sacks and Kucharik 2011). Such effects may not be linear because as air temperatures increase above the optimum, instead of falling at a rate linearly with the temperature increase, yield decreases sharply. This effect was demonstrated in multiple crops in a recent analysis by Schlenker and Roberts (2009) where the yield for corn, soybean, and cotton was predicted to gradually increase with temperatures up to 29°C (corn), 30°C (soybean), and 32°C (cotton) before decreasing sharply with temperature increases beyond this threshold.

The impact of temperature change on predicted crop yield into the remainder of this century differs between studies. Simple crop simulation models have predicted that continued increases in temperature will cause yield declines between 2.5% and 10% across a number of agronomic species (Hatfield et al. 2011). Lobell et al. (2011) produced estimates of yield decline between 3.8% and 5%, while Schlenker and Roberts (2009) used a statistical approach to predict wheat, corn, and cotton yield declines of 36–40% under a low emissions scenario and between 63% and 70% for high emissions scenarios. Their approach did not, however, consider the potential interacting effects of rising atmospheric CO_2 on crop growth, variation among crop genetics, the effect of pests on crop yield, or the use of adaptive management strategies, for example, fertilizers, rotations, tillage, or irrigation.

In grain sorghum, high temperature can result in significant increases in leaf numbers, particularly when reproductive development is arrested without large decrease in leaf photosynthetic rates (Prasad et al. 2006a). The length of time from floral initiation (panicle initiation) to anthesis (panicle exsertion) is generally decreased by moderate increases in temperature, but severe high-temperature stress delays panicle initiation, restricts panicle exsertion, and potentially delays flowering (Prasad et al. 2006a).

Early reproductive processes such as micro- and megasporogenesis, anthesis, pollination, pollen tube growth, fertilization, and seed-set development are all highly susceptible to increased temperature. Failure of any of these processes can decrease fertilization or increase early embryo abortion, leading to lower seed/grain production, thereby limiting crop yield. High-temperature stress prior to or at anthesis causes a significant increase in floral abortion and lower seed numbers in many grain crops (Prasad et al. 1999a, b, 2008a). Exposure to high-temperature stress during flowering results in pollen sterility and loss of seed-set in legumes (groundnut, Prasad et al. 2000a, b, 2003; dry bean, Prasad et al. 2002; soybean, Djanaguiraman et al. 2013a, b) and cereals (rice, Jagadish et al. 2007; wheat, Saini et al. 1983; Prasad et al. 2008b). Exposure to temperatures >37°C for as short as 1 h during flowering has been shown to decrease seed-set in rice (Matsui et al. 2001). Similarly, exposure to temperatures >33°C for 6 h after anthesis decreased seed-set in peanut (Prasad et al. 2000b). The timing of high-temperature exposure relative to peak flowering is therefore critical (Wheeler et al. 2000). Furthermore, high-temperature stress during flowering can reduce pollen germination by degrading tapetal cells. Recent studies have shown that high-temperature stress results in increased reactive oxygen species production, membrane damage, and changes in phospholipid profiles of pollen, leading to lower pollen viability and germination (Prasad and Djanaguiraman 2011; Djanaguiraman et al. 2013a, b).

Prolonged high-temperature stress shortens the reproductive development duration (period during which potential kernel or seed numbers are determined) and the grain filling duration (during which the grain or seed weight is determined), leading to smaller seed size. For most crop species, particularly those for which there is a physical restriction for growth of seeds such as rice (which has fixed pericarp) and legumes (which have fixed locule size for development of the seed), yield capacity is mainly a function of seed numbers per unit area and seed filling duration. Although there is often a slight increase in seed filling rate under high temperature, this does not compensate for the reduction in seed filling duration.

Global climate models predict that nighttime temperatures are expected to increase at a faster rate than daytime temperatures. High nighttime temperature has been shown to have a more pronounced negative effect than daytime temperatures on the yield of rice, with a 10% reduction in yield for every 1°C increase in nighttime temperature. This study recorded an overall increase in mean nighttime temperature of 1.13°C from 1992 to 2003 (Peng et al. 2004). Studies have also shown that high nighttime temperature decreases pollen viability, spikelet fertility, and grain weight in rice (Morita et al. 2002; Mohammed and Tarpley 2009). Wheat yields have been observed to decrease linearly with increasing nighttime temperatures. When compared with controls at 14°C, grain fill duration decreased by 3 and 7 days, respectively, at night temperatures of 20°C and 23°C (Prasad et al. 2008b). Nighttime temperature >20°C additionally decreased spikelet fertility, grains per spike, and grain size. Decreases in photosynthesis were observed at night temperatures >14°C. High nighttime temperatures significantly decreased yields in soybean and grain sorghum by both increasing respiration rate and decreasing the photosynthesis rate (Prasad and Djanaguiraman 2011; Djanaguiraman et al. 2013a, b).

8.5 The interaction of climate change factors on crop development

Better understanding of the interactions between climate change factors is critical to understanding response of crops to changing environments. In particular, the interactions between elevated temperature and elevated CO_2, water, and nutrients should be explored.

8.5.1 The interaction of rising temperature and CO_2

Together with rising temperatures, atmospheric CO_2 concentrations are increasing with global climate change (IPCC 2007b). At low light, plant growth is strongly carbon limited and thus responds better to high CO_2.

Elevated CO_2 decreases stomatal conductance leading to decreased water loss, lower total water use, and increased water use efficiency. Thus, under water stress (drought) conditions, crops will survive longer under elevated CO_2. Water savings from lower transpiration will be offset by larger leaf areas and higher tissue temperatures in these conditions. Furthermore, the response of plants to elevated CO_2 is generally reduced at low levels of plant nutrition. An increase in the carbohydrate pool due to increased photosynthesis can only stimulate growth if crops can acquire more nutrients and use them more efficiently.

The beneficial effects of elevated CO_2 on growth and yield decrease at supraoptimal temperatures (Prasad et al. 2002, 2003, 2006a). Small increases in air temperature would more than offset the water-saving effects of CO_2 from decreased stomatal conductance. Studies using rice, dry bean, and sorghum have shown that plants grown at elevated CO_2 displayed decreases in optimum and ceiling temperatures for seed-set by about 2°C (Matsui et al. 1997; Prasad et al. 2002, 2006a).

8.5.2 The interaction of high-temperature and drought stress

High-temperature stress and drought stress commonly occur in combination under field conditions. The effects of drought on leaf water potential are progressive. When leaf water potential is low, stomata close, decreasing transpiration. Leaf water potential is also dependent upon vapor pressure deficit (discussed in Section 8.3). High temperature increases evaporation from the soil surface and transpiration from the leaf surface, due to higher vapor pressure deficits. In this way, high temperatures interact strongly with drought and exacerbate its effects when water is withheld (Machado and Paulsen 2001).

The simultaneous effects of high-temperature and drought stress on crop performance and yield may be different than the individual stresses alone. In wheat, the combined effect of high temperature and drought on leaf chlorophyll content, grain-set, seed yield, and harvest index was more severe than the additive effects of these individual stresses (Prasad et al. 2011; Pradhan et al. 2012a). Differential effects between traits were, however, observed. Grain number was more sensitive to high-temperature stress, while grain weight displayed greater sensitivity to drought. This research highlights the differential interactive effects between high temperature and drought on various physiological and yield processes. There is a need to accurately model these interactive effects on crops so that realistic assessments of impacts and adaptive strategies can be made (Nuttall et al. 2012a). Such adaptation will involve genetic tolerance to very high temperatures under both irrigated and drought conditions.

8.6 The impact of global climate change on food quality and plant nutrient demand

Because CO_2 is an essential substrate for photosynthesis, early modeling analyses assumed that the fertilizing effects of enhanced CO_2 would compensate for high-temperature-induced growth decline, predicting that the major uncertainty of our changing climate would relate to precipitation (Wang and Connor, 1996; O'Leary and Anwar, 2008). While this is generally true, the combined effects of elevated temperature, CO_2, and nutrient demand need to be addressed (Leakey et al. 2009). Elevated CO_2 raises the yield of C3 crops and increases nutrient uptake but results in lower quality biomass and grain (Amthor 2001; Hogy et al. 2009; Tausz et al. 2013). Grain quality is differentially affected by different climatic components. For example, the proportion of flour extracted from whole wheat grain (milling yield) has been shown to be unaffected by elevated CO_2, but baking quality is strongly reduced (Nuttall et al. 2012b). Not only are the supplies of staple foods at risk from rising temperatures, but combined with higher atmospheric CO_2 and lower precipitation in many parts of the world, the resultant decrease in food quality (hidden hunger) is expected to become more problematic.

The enhanced biomass and yield of C3 crops resulting from the CO_2 fertilization effect (Ainsworth et al. 2008) will significantly increase nutrient demand, questioning the value of low-input agriculture as a way to manage food production in the face of climate change (Connor 2008; Connor 2013). A focus is therefore needed on how to boost nutrient supplies, rather than reduce them. New genetic and agronomic technology has always offered hope of quantum leaps in productivity and profitability (Cassman 2007), but slow gains to date raise doubt that the present productivity gains will be sufficient to feed the projected future demands (Fischer 2011). A focus on nutrients and their supply and management, including pests and disease management, will be crucial areas of future agricultural research. This will ultimately require an integrated systems approach through modeling and targeted experimental research.

8.7 Breeding high-temperature stress tolerance using crop wild relatives

There is limited variability among cultivar or breeding material in most grain crops. Genetic variation for high-temperature tolerance does, however, exist within land races and wild species. High-temperature tolerance during the vegetative stage was shown to be higher in wild wheat relatives of *Aegilops speltoides* and *Aegilops tauschii* than in *Triticum urartu* and *Triticum boeoticum* (Waines 1994). A high level of tolerance also was observed in the line *Triticum dicoccoides* 600808 from Jordan (Rekika et al. 1997). Pradhan et al. (2012b) evaluated 52 accessions belonging to 5 *Aegilops* species at optimum temperature (25/19°C day/night) and high temperature (36/30°C). Stress was imposed at anthesis and continued for 16 days. Across wheat species, high temperature decreased chlorophyll content, grain number per spike, individual grain weight, and grain yield per plant by 38, 40, 56, and 70%, respectively. *A. speltoides* and *Aegilops geniculata* displayed greater tolerance to high temperature for yield (58–61% decline from optimum temperature), whereas *Aegilops longissima* was highly susceptible (84% decline). Yield tolerance to high temperature was associated with greater grain number per spike and/or individual grain weight. Within *A. speltoides*, accession TA 2348 displayed significant tolerance to high temperature, with a 13.5% decline in grain yield and a high-temperature susceptibility index (HSI) of 0.23 in contrast to highly susceptible accessions displaying a >82% yield decline and HSI >1.4. Among *A. geniculata* accessions, two were moderately high-temperature tolerant, TA 2899 and TA 1819, with an HSI of 0.80. These results suggest that genetic variability among *Aegilops* species can be utilized for improving high-temperature tolerance in wheat during the reproductive stages of crop development (Pradhan et al. 2012b).

In rice, the daily timing of flowering can be used as a mechanism to avoid high temperatures during anthesis, pollen germination, and pollen tube growth. Genotypes are known to differ in this trait (Prasad et al. 2006b; Jagadish et al. 2007). Jagadish et al. (2007) and Ishimaru et al. (2010) reported that early morning flowering (EMF) could be a viable strategy to overcome the adverse effect of high-temperature stress during the later hours of the morning but identified few *O. sativa* accessions displaying this response. Screening of wild rice accessions at the International Rice Research Institute (IRRI) revealed the EMF trait in *Oryza minuta* and *Oryza officinalis*. The latter was successfully used to introgress the EMF trait into the modern variety Koshihikari (*O. sativa*), accelerating flowering by as much as 4 h (Ishimaru et al. 2010). This shift in flowering time toward the cooler early morning resulted in significantly lower high-temperature-induced spikelet sterility. Scafaro et al. (2010) have additionally reported that *Oryza meridionalis*, a wild relative of *O. sativa*, displays greater high-temperature tolerance than *O. sativa*, as shown by comparing leaf elongation, photosynthetic rates, and accumulation of heat-induced proteins.

8.8 Global food production and food security

Climate change severity will be reinforced by the unprecedented increase in the global population which is expected to reach nine billion by 2050, with further increases through 2100. Each new arrival will add to humanity's expanding ecological footprint, as each person now lives longer, consumes food above the subsistence level for 90% of the population, and is more demanding of both energy sources and manufactured goods. It is a scenario that will challenge agriculture to at least double food production this century, despite increasing urban competition for land and water (Redden et al. 2011).

In this overview, the current status of the basic starch-rich staple foods and high-protein plant foods of the world is assessed and the impact climate uncertainty might have on their production. Animal foods will not be discussed here, except to note that production of feed for animals competes with food production for humans and that increased consumption of meat has a higher income elasticity of demand as incomes increase. The world production of the basic staple foods, wheat, rice, maize, potato, cassava, and sorghum, includes both developed and developing countries. The majority of demand is usually met by local production as only about 5–10% of global staple crop production is traded, although certain countries are major exporters such as the USA, Australia, and Canada for 50–70% of the wheat production (GIEWS 2011).

8.8.1 Wheat production

The leading countries in wheat production (653 million metric tonnes (m.mt)) in 2010/2011 were the European Union (EU) (132 m.mt), China (126), India (91), the USA (61), Russian Federation (56), Canada (27), Australia (26), and Pakistan (23) followed by Brazil, Turkey, Kazakhstan, Iran, Ukraine, and Argentina (IGC 2012). Principal importers were Japan, South Korea, Egypt, Saudi Arabia, Iran, Iraq, and Mexico; principal exporters were the USA, Canada, Australia, the EU, Russia, Ukraine, and Kazakhstan.

8.8.2 Rice production

World rice production in 2010 was estimated at 700 m.mt corresponding to 466 m.mt of milled rice, led by China (200), India (141), Indonesia (66), Bangladesh (51),Vietnam (40), Thailand (32), Myanmar (31), Philippines (16), Brazil (13), the USA (11), and Japan (11), followed by other countries in South and East Asia, West Africa, and Latin America (FAO 2011). About 7.3% of world production is internationally traded, principal exporters being Thailand, Vietnam, the USA, India, Pakistan, and Cambodia, while major importers were Bangladesh, Nigeria, Philippines, Malaysia, China, North Korea, South Africa, and Ivory Coast (FAO 2011). Production was distorted by floods in South and Southeast Asia in 2010 and 2011.

8.8.3 Potato production

Potato production in 2009 was 330 m.mt, mainly comprising China (75), India (37), Russia (21), Ukraine (19), the USA (18), Germany (10), Poland (9), Bangladesh (8), Belarus (8), France (7), Netherlands (7), and Belarus

(7) (Geohive 2010). The remaining third of production includes the Andes region where potato originated. Food use of potato represents two-thirds of total production, while animal feed use amounts to one-third.

8.8.4 Maize production

Total world maize production was 866 m.mt in 2010/2011 (IGC 2012), with about 11% being traded mainly for feed as well as industrial consumption including ethanol. The USA is a leading producer with 350 m.mt projected for 2012, while in China maize production (163 m.mt) now exceeds that of wheat driven by the expanding demand for feed. Other significant producers are Brazil (51), India (17), Mexico (20), and Argentina (13). Productivity varies from 2 t ha^{-1} in India to 10 t ha^{-1} in the USA. The USA is the dominant exporter, followed by Argentina, China, Hungary, and South Africa (white maize), while Japan, South Korea, Taiwan, Mexico, Malaysia, and the EU import maize (FAO 2012a).

8.8.5 Sorghum production

World sorghum production was projected at 65 m.mt in 2010, with Nigeria, the USA, and India being major producers with 7.5–10 m.mt, followed by Mexico, Sudan, Argentina, and Australia (Agrostats 2009). Sorghum is an important food in the West African Savannah/Sahel, Ethiopia, and Somalia. There is a wide disparity in yield from 0.8 t ha^{-1} in Africa to 4 t ha^{-1} in North America. Less than 5% of total production is traded, mostly for feed, to Japan, Central Asia, Mexico, Caribbean, Sudan, East Africa, and Niger. Per capita consumption is 90–100 kg $year^{-1}$ in Burkina Faso and Sudan and 70 kg $year^{-1}$ in India, as a food security crop (FAO and ICRISAT 1966).

8.8.6 Cassava production

World cassava production was 91 m.mt in 2011, led by Brazil (27), Indonesia (11), Nigeria (10), Zaire (9), India (6), and Mozambique (2.5) (FAO 2012b). Brazil and Thailand are major exporters to the EU for starch. Cassava is a drought-resilient crop, with per capita consumption above 200 kg $year^{-1}$ in Africa, and is also regionally important in Latin America and the Caribbean.

8.8.7 Pulse production

Pulses, with a world production of 56 m.mt in 2007 (Tata Strategic Management Group 2012), provide an important source of high-protein food in developing countries with diets that consist primarily of high-carbohydrate staple foods. Ninety-five percent of pulses are cultivated in the developing world. The principal pulses are *Phaseolus* beans (46%), chickpea (22%), faba bean (10%), and 7% each for lentil, pigeon pea, and cowpea (SPIA Draft Report 2011). India is the leading producer of *Phaseolus* beans, chickpea, pigeon pea, and lentil. Other important producers are Brazil, Burma, Mexico, and Tanzania for *Phaseolus* beans; Pakistan, Iran, Turkey, Myanmar, and Ethiopia for chickpea; Myanmar, Kenya, and Malawi for pigeon pea; and Turkey, Iran, Nepal, Syria, Bangladesh, and Ethiopia for lentil. China is the leading faba bean producer followed by Ethiopia, Morocco, Egypt, Sudan, and Tunisia. The major cowpea production is in West Africa, led by Niger and Nigeria, followed by Burkina Faso, Mali, and Senegal. Soybeans account for only one percent of global pulse production with 0.7.m.mt, mainly produced in the USA, Brazil, and Argentina; and China is the principal importer.

8.8.8 Predicted impacts of climate change on global crop production

As described earlier, major climate change impacts by 2030 are expected for maize with a 30% yield reduction in South Africa and reductions in South and Southeast Asia as well as China (Lobell et al. 2008). Wheat production will be reduced in these regions, by up to 5% in South Asia with a 75% probability of occurrence. The situation will be similar for rice, millet, and *Brassica* crops. The dimension of these impacts could be very severe in India, where the population is expected to reach 1.75–1.82 billion in 2051 (Population Reference Bureau 2007) – a net result of less food per capita.

The impact of climate change on food production in other regions will be crop specific. Brazil and the Andean region are predicted to have reduced wheat and rice production; Brazil may also experience reduced production of maize, cassava, and soybean. The West African Sahel is expected to have more desert encroachment, with reduced production of sorghum, although millet and cowpea production may rise. In tropical West Africa, yields of peanuts, yams, and cassava are likely to decline. Central Africa may see reduced production of both sorghum and millet. East Africa may have an increase in yield for barley but a reduction for cowpea. Growing seasons will be more extreme in India and Pakistan.

8.9 Crop nutritional content

One index of worldwide nutrition status is the daily caloric intake per capita (kcal). Many countries are still below the 2500 kcal recommended minimum daily intake, and these include (with national daily intake bracketed)

Antigua and Barbuda (2228), Armenia (2447), Bangladesh (2284), Bolivia (2105), Cambodia (2107), Comoros (2212), North Korea (2149), Dominican Republic (2319), Ecuador (2228), Georgia (2435), Grenada (2375), Guatemala (2188), Haiti (1848), Honduras (2493), India (2278), Indonesia (2464), Lao PDR (2149), Madagascar (2097), Mongolia (2138), Myanmar (2230), Nepal (2281), Nicaragua (2194), Niger (2170), Pakistan (2361), Panama (2398), Peru (2334), Philippines (2412), Seychelles (2405), Solomon Islands (2369), Sri Lanka (2320), Suriname (2469), Tajikistan (2027), Timor-Leste (2011), Vietnam (2311), Yemen (2061), and sub-Saharan Africa (except Nigeria, Ivory Coast, and Republic of South Africa) (FAOSTAT 2012). Food supplies are expected to be critically low in South and Southeast Asia and sub-Saharan Africa, which are among the regions most vulnerable to climate change.

An increased production of staple crops is needed to meet the demand for caloric intake of a growing world population and a desired increase in daily energy availability from about 2770 kcal in 2003–2005 to an average of 2970 kcal in developing countries by 2050 (FAO 2009). However, there is a clear need to reorient agricultural production from a mere increase in food quantity toward delivery of a better quality of food. Poverty-induced restrictive diets mainly based on staple crops, like maize, rice, or cassava, result in prolonged dietary deficits in micronutrients and vitamins and associated diet-related deadly diseases. In the coming decades, the lack of micronutrients in diets may become a much more relevant nutritional issue in sub-Saharan Africa and South Asia than simply increasing the staple food supply and the amount of calories consumed. As incomes rise, overnourishment from imbalanced diets also becomes a matter for concern. The quality of food is a major decisive factor to address malnutrition in both of these situations.

Biofortification of staple crops may offer a partial solution to improve the content of essential nutrients in the diets of malnourished people (Bouis et al. 2011). A classic example is the development of 'Golden Rice' lines with enhanced levels of β-carotene in the endosperm so that a single serving of Golden Rice can supply the daily recommended intake of provitamin A to alleviate vitamin A deficiency symptoms (Paine et al. 2005). A much easier solution to obtain a balanced and more diverse diet is the production and consumption of a wider range of food crops, especially vegetables and fruits. Diversifying agricultural production will help to increase resilience of farming systems as the global climate changes, especially among smallholder farmers. Efficient adaptation strategies for a changing climate will require, among other measures, the preservation of the remaining biodiversity, both *in situ* as well as in gene banks, and facilitate access to genetic resources of crops and their wild relatives (HPLE 2012).

8.10 Discussion

Accelerated climate change is expected to have a significant, but variable, impact on the world's major cropping zones. Crops will experience increasingly warmer, drier, and more variable growing conditions in the temperate to subtropical latitudes toward 2050 and beyond. However, short-term 1–5 day spikes of extreme weather may be hotter than previously experienced, occasional frosts in the reproductive period more frequent, and rainfall less dependable and more variable. In contrast, more irregular and concentrated storms may increase flood damage to crops in the tropics. These predicted trends will create a production and food risk for subsistence farmers and for increasingly urbanized societies worldwide.

There may also be new growing opportunities at higher latitudes and altitudes, with warmer conditions extending the length of growing seasons. A carbon fertilizer boost should additionally enhance photosynthetic efficiency in C3 plants, in response to higher CO_2 levels. Such gains are, however, likely to be more than offset by reduced production due to accompanying shorter crop cycles in the warmer temperate–tropical zones and desertification in marginal zones such as the West African Sahel, upland South Africa, and the Deccan plateau of central India.

Rainfed marginal zones are at risk, and also some irrigated zones where ground water is depleted in conjunction with a reduced recharging of aquifers. Survival strategies will need to consider whole-farm water conservation measures, modified or reduced tillage, and selective cropping of the deeper soils and topographies that benefit from water runoff. Complementary breeding objectives could include tolerance of extreme high temperatures in the reproductive period, drought avoidance traits such as altered maturity and waxy and hirsute leaves, and drought tolerance expressed through stay-green leaves and deeper, as well as more opportunistic, surface root systems.

Crop choices for these more adverse conditions may require a shift to more heat- and drought-tolerant crops, possibly with shorter growing seasons, such as finger millet in the Sahel or very early maturing hybrid pigeon pea in East Africa. A reduction in the toxicity of *Lathyrus* (grass pea) has potential benefit for this famine relief crop, which is well known for tolerance to drought, salinity, and waterlogging. Currently overlooked minor crops and food sources such as indigenous vegetables and fruits may rise to prominence as locally important crops, to reduce reliance on the major staple foods of wheat, rice, maize, and potato. Diversifying agricultural production and food sources not only will help to increase resilience of farming systems as the global climate changes, especially for smallholder farmers, but will also help to overcome malnutrition afflicting millions of people worldwide. This tendency will gain strength as a more affluent middle class in population-dense countries such as China and India creates steadily increasing demand for dietary diversity.

Humans need a balanced diet of carbohydrates, proteins, and vitamins sourced from plants directly as food and indirectly as animal feed. Society generally will consume more animal/fish produce when available and affordable, with a high income elasticity of demand. How will this surge in demand for animal and fish produce be sustained in the face of climate change? Already maize for pig feed has displaced rice as the premier crop in China, along with amalgamation of small farms to create large holdings suited to mechanization and high-input agriculture. In a similar way, urbanization is a driver for change in the agricultural output of most countries in South, East, and Tropical Asia. Currently Indonesia has great difficulties in meeting the demand for beef, even with high levels of importation, due in part to sudden export restrictions from Australia motivated by concern for animal welfare. Innovation in sources of feed will be needed, such as algae, with greater photosynthetic efficiency and higher biomass production than higher plants (Wellinger 2007).

To maintain and increase crop productivity in increasingly hostile environments, novel sources of genetic variation must be sought if crops are to adapt to climate change. There are large but little explored gene pools among the crop wild relatives, some readily available to the domestic gene pools through conventional hybridization, while the more genetically distant relatives in the tertiary gene pool will require genetic engineering techniques for introgression of the desired major and quantitative genes. For traits that are difficult to phenotype, marker-assisted selection will be important, especially within-gene or perfect markers. Often crop wild relatives have survived on nonarable and marginal lands in stressful environments. Because only a small part of this wide diversity was captured during the domestication of crops, there is a real likelihood that tolerance of extreme temperature, drought, and soil stresses might be found in the wild gene pools, with an evolutionary history extending back hundreds of millennia and across very diverse environments. In contrast, the evolution of crops is relatively recent, comprising the last 1000–11,000 years. Exploitation of genetic resources for tolerance of warmer global temperatures and associated extreme heat spikes during crop seasons will primarily depend on domestic landraces from extreme environments, followed by the selective filtering of desired genes from crop wild relatives.

8.11 Conclusions

There is an urgency to focus agricultural research on various aspects that high global temperature presents in respect to food and nutrition security. This includes dealing with the interactive effects of the other three primary climate factors of water, nutrients, and CO_2. While efforts need to continue on major genetic and agronomic fronts, it is clear that without more investment there is a real risk of catastrophic failure in global food and nutrition security. It is essential to maintain a cooperative focus to achieve a sufficient level of nutritious food for all.

References

Agrostats (2009) World sorghum production. http://www.agrostats.com/world-statistic/world-sorghum-production.html (accessed on September 13, 2013).

Ainsworth EA, Beier C, Calfapietra C, Ceulemans R, Tardifs MD, Farquhar GD, Godbold DL, Hendrey GR, Hickler T, Kaduk J, Karnosky D, Kimball BA, Korner C, Koornneef M, Lufurgc T, Leakey ADB, Lewin KF, Long SP, Manderscheid R, McNeil DL, Mies TA, Miglietta F, Morgan JA, Nagy J, Norby RJ, Norton RN, Percy KE, Rogers A, Soussana J-F, Stitt M, Weigel H-J, White J (2008) Next generation of elevated [CO_2] experiments with crops: A critical investment for feeding the future world. Plant Cell Environ. 31: 1317–1324.

Alfaro EJ, Gershunov A, Cayan D (2006) Prediction of summer maximum and minimum temperature over the central and western United States: The roles of soil moisture and sea surface temperature. J. Climate 19: 1407–1421.

Amthor JS (2001) Effects of atmospheric CO_2 concentration on wheat yield: Review of results from experiments using various approaches to control CO_2 concentration. Field Crops Res. 73: 1–34.

Bouis HE, Hotz C, McClafferty B, Meenakshi JV, Pfeiffer H (2011) Biofortification: A new tool to reduce micronutrient malnutrition. Food Nutr. Bull. 32: S31–S40.

Buck AL (1981) New equations for computing vapor pressure and enhancement factor. J. Appl. Meterol. 20: 1527–1532.

Burke JJ, Mahan JR, Hatfield JL (1988) Crop-specific thermal kinetic windows in relation to wheat and cotton biomass production. Agron. J. 80: 553–556.

Burney JA, Davis SJ, Lobell DB (2010) Greenhouse gas mitigation by agricultural intensification. Proc. Natl. Acad. Sci. USA 107: 12052–12057.

Cassman KG (2007) Editorial response: Can organic agriculture feed the world – Science to the rescue? Renew. Agric. Food Syst. 22: 83–84.

Conaty WC, Burke JJ, Mahan JR, Nielsen E, Sutton BG (2012) Determining the optimum plant temperature of cotton physiology and yield to improve plant-based irrigation scheduling. Crop Sci. 52: 1828–1836.

Connor DJ (2008) Organic agriculture cannot feed the world. Field Crops Res. 106: 187–190.

Connor DJ (2013) Organically grown crops do not a cropping system make and nor can organic agriculture nearly feed the world. Field Crops Res. 144: 145–147.

Djanaguiraman M, Prasad PVV, Schapaugh WT (2013a) High day- or nighttime temperature alters leaf assimilation, reproductive success, and phosphatidic acid of pollen grain in soybean (Glycine max L. Merr.). Crop Sci. 53: 1594–1604.

Djanaguiraman M, Prasad PVV, Boyle DL, Schapaugh WT (2013b) Soybean pollen anatomy, viability, and pod set under high temperature stress. J. Agron. Crop Sci. 199: 171–177.

Food and Agriculture Organization (FAO) (2004) Impact of climate change on agriculture in Asia and the Pacific. Twenty-seventh FAO Regional Conference for Asia and the Pacific, May 17–21, 2004. Beijing, China.

FAO (2009) Global agriculture towards 2050. High level expert forum – How to feed the world in 2050. Rome, Italy, October 12–13, 2009. http://www.fao.org/fileadmin/templates/wsfs/docs/Issues_papers/HLEF2050_Global_Agriculture.pdf (accessed on August 20, 2013).

FAO (2011) Rice market monitor. http://www.fao.org/economic/est/publications/rice-publications/rice-market-monitor-rmm/en/ (accessed on August 20, 2013).

FAO (2012a). White maize: A traditional food in developing countries. International trade. http://www.fao.org/docrep/W2698E/W2698E00.htm (accessed on August 20, 2013).

FAO (2012b). World production and trade of cassava products. http://www.fao.org/docrep/x5032e/x5032E08.htm (accessed on August 20, 2013).

FAOSTAT (2012) Food balance sheets. http://faostat3.fao.org/home/index.html#VISUALIZE_BY_DOMAIN (accessed on August 20, 2013).

FAO and ICRISAT (1966) The world sorghum and millet economies. Facts, trends and outlook. ICRISAT/FAO, India/Italy. 68pp.

Fischer RA (2011) Wheat physiology: A review of recent developments. Crop Pasture Sci. 62: 95–114.

Geohive (2010) http://www.geohive.com/charts/ag_potato.aspx (accessed on August 20, 2013).

GIEWS (2011) Food outlook. Global market analysis. FAO November 2011. http://www.fao.org/giews/english/fo/index.htm (accessed on August 20, 2013).

Hatfield JL, Prueger JH (2011). Agroecology: Implications for plant response to climate change. In: Yadav SS, Redden RJ, Hatfield JL, Lotze-Campen H, Hall AE (eds). Crop adaptation to climate change. John Wiley & Sons, Ltd, Chichester.

Hatfield JL, Gitelson AA, Schepers JS, Walthall CL (2008) Application of spectral remote sensing for agronomic decisions. Agron J. 100: S117–S131.

Hatfield JL, Boote KJ, Kimball BA, Ziska LH, Izaurralde RC, Ort D, Thomson AM, Wolfe DW (2011) Climate impacts on agriculture: Implications for crop production. Agron. J. 103: 351–370.

Hogy P, Wieser H, Kohler P, Schwadorf K, Breuer J, Franzaring J, Muntifering R, Frangmeier A (2009) Effects of elevated CO_2 on grain yield and quality of wheat: Results from a 3-year free air CO_2 enrichment experiment. Plant Biol. 11: 60–69.

HLPE (2012) Climate change and food security. A report by the High Level Panel of Experts on Food Security and Nutrition of the Committee on World Food Security, Rome, Italy.

IGC Grains Conference (2012) Grain market report. http://www.igc.int/downloads/gmrsummary/gmrsumme.pdf (accessed on August 20, 2013).

Intergovernmental Panel on Climate Change (IPCC) (2007a) The physical science basis. In: Solomon S, Qin D, Manning M, Chen Z, Marquis M, Averyt M, Tignor M, Miller M (eds). Contribution of Working Group I to the Fourth Assessment Report of the Intergovernmental Panel on Climate Change. Cambridge University Press, Cambridge.

IPCC (2007b) Impacts, adaptation and vulnerability. In: Parry L, Canziani OF, Palutikof JP, van der Linden PJ, Hanson CE (eds). Contribution of Working Group II to the Fourth Assessment Report of the Intergovernmental Panel on Climate Change. Cambridge University Press, Cambridge.

IPCC (2012) Managing the risks of extreme events and disasters to advance climate change adaptation. A special report of Working Groups I and II of the Intergovernmental Panel on Climate Change In: Field CB, Barros V, Stocker TF, Qin D, Dokken DJ, Ebi KL, Mastrandrea MD, Mach KJ, Plattner G-K, Allen SK, Tignor M, Midgley PM (eds). Cambridge University Press, Cambridge, New York, 582pp.

Ishimaru T, Hirabayashi H, Ida M, Takai T, San-Oh YA, Yoshinaga S, Ando I, Ogawa T, Kondo M (2010) A genetic resource for early-morning flowering trait of wild rice *Oryza officinalis* to mitigate high temperature-induced spikelet sterility at anthesis. Ann. Bot. 106: 515–520.

Jagadish SVK, Craufurd PQ, Wheeler TR (2007) High temperature stress and spikelet fertility in rice (*Oryza sativa* L.). J. Exp. Bot. 58: 1627–1635.

Karl, TR, JM Melillo, TC Peterson (2009) Global climate change impacts in the United States. Cambridge University Press, Cambridge.

Katz RW, Brown BG (1992) Extreme events in a changing climate: Variability is more important than averages, Clim. Change 21: 289–302.

Klein JA, Harte J, Zhao XQ (2007) Experimental warming, not grazing, decreases rangeland quality on the Tibetan plateau. Ecol. Appl. 17: 541–557.

Knowles N, Dettinger MD, Cayan DR (2006) Trends in snowfall versus rainfall in the western United States. J. Climate 19: 4545–4559.

Leakey ABD, Ainsworth EA, Bernacchi CJ, Rogers A, Long SP, Ort DRT (2009) Elevated CO_2 effects on plant carbon, nitrogen, and water relations: Six important lessons from FACE. J. Exp. Bot. 60: 2859–2876.

Lobell DB, Burke MB, Tebaldi C, Mastrandrea MD, Falcon WP, Naylor RL (2008) Prioritizing climate change adaptation needs for food security in 2030. Science, 319: 607–610.

Lobell DB, Schlenker W, Costa-Roberts J (2011) Climate trends and global crop production since 1980. Science 333: 616–620.

Machado S, Paulsen GM (2001) Combined effects of drought and high temperature on water relations of wheat and sorghum. Plant Soil 233: 179–187.

Matsui T, Namuco OS, Ziska LH, Horie T (1997) Effect of high temperature and CO_2 concentrations on spike sterility in indica rice. Field Crops Res. 51: 213–219.

Matsui T, Omasa K, Horie T (2001) The difference in sterility due to high temperature during flowering period among japonica rice varieties. Plant Prod. Sci. 4: 90–93.

Meehl GA, Stocker TF, Collins WD, Friedlingstein P, Gaye AT, Gregory JM, Kitoh A, Knutti R, Murphy JM, Noda A, Raper SCB, Watterson IG, Weaver AJ, Zhao Z-C (2007) Global climate projections. In: Solomon, S, Qin D, Manning M, Chen Z, Marquis M, Averyt KB, Tignor M, Miller HL (eds). Climate change 2007: The physical science basis. Contribution of Working Group I to the Fourth Assessment Report of the Intergovernmental Panel on Climate Change. Cambridge University Press, Cambridge.

Mohammed AR, Tarpley L (2009) Impact of high nighttime temperature on respiration, membrane stability, antioxidant capacity, and yield of rice plants. Crop Sci. 49: 313–322.

Morita S, Shiratsuchi H, Takanashi J, Fujita K (2002) Effect of high temperature on ripening in rice plants – Comparison of the effects of high night temperatures and high day temperatures. Jpn. J. Crop Sci. 71: 102–109.

Müller C, Cramer W, Hare WL, Lotze-Campen H (2011) Climate change risks for African agriculture. Proc. Natl. Acad. Sci. USA 108: 4313–4315.

Nelson GC, Rosegrant MW, Koo J, Robertson R, Sulser T, Zhu T, Ringler C (2009) Climate change: Impact on agriculture and costs of adaptation. International Food Policy Research Institute, Washington, DC. www.ifpri.org/publication/climate-change-1 (accessed on August 20, 2013).

Nelson GC, Rosegrant MW, Palazzo A, Gray I, Ingersoll C, Robertson R, Tokgoz S (2010) Food security, farming, and climate change to 2050: Scenarios, results, policy options. International Food Policy Research Institute, Washington, DC.

Nuttall J, Brady S, Brand J, O'Leary G, Fitzgerald GJ (2012a) Heat waves and wheat growth under a future climate. In: Yunusa I (ed.). Capturing opportunities and overcoming obstacles in Australian agronomy. Proceedings of 16th Australian Agronomy Conference 2012, 14–18 October 2012, Armidale, NSW, Australia. http://www.agronomy.org.au. http://www.regional.org.au/au/asa/2012/climate-change/8085_nuttalljg.htm (accessed on August 20, 2013).

Nuttall J, O'Leary G, Panozzo J, Fitzgerald GJ (2012b) A model for predicting milling yield in wheat. In: Yunusa I (ed.). Capturing opportunities and overcoming obstacles in Australian agronomy. Proceedings of 16th Australian Agronomy Conference 2012, 14–18 October 2012, Armidale, NSW, Australia. http://www.agronomy.org.au. http://www.regional.org.au/au/asa/2012/breeding/8082_nuttalljg.htm (accessed on August 20, 2013).

O'Leary G, Anwar M (2008) Simulation of wheat grain yield under elevated CO_2 conditions in the Victorian Wimmera. In: Unkovich MJ (ed.). Global issues, paddock action. Proceedings of the 14th Australian Agronomy Conference, 10–14 September 2008, Adelaide, South Australia. Australian Society of Agronomy. http://www.agronomy.org.au (accessed on August 20, 2013).

O'Leary GJ, Connor DJ, White DH (1985) Effect of sowing time on growth, yield and water use of rain fed wheat in the Wimmera, Vic. Aust. J. Agr. Res. 36: 187–196.

O'Leary G, Christy B, Weeks A, Nuttall J, Riffkin P, Beverly C, Fitzgerald G (2011) Downscaling global climatic predictions to the regional level: A case study of regional effects of climate change on wheat crop production in Victoria, Australia. In: Yadav SS, Redden RJ, Hatfield JL, Lotze-Campen H, Hall AE. Crop adaptation to climate change (1st ed.). John Wiley & Sons, Ltd, Chichester.

Paine JA, Shipton CA, Chaggar S, Howells RM, Kennedy MJ, Vernon G, Wright SY, Hinchliffe E, Adams JL, Silverstone AL, Drake R (2005) Improving the nutritional value of golden rice through increased pro-vitamin A content. Nat. Biotechnol. 23: 482–487.

de la Peña RC, Ebert AW, Gniffke P, Hanson P, Symonds RC (2011) Genetic adjustment to changing climates: Vegetables. In: Yadav SS, Redden RJ, Hatfield JL, Lotze-Campen H, Hall AE (eds). Crop adaptation to climate change (1st ed.) John Wiley & Sons, Ltd, Chichester.

Peng S, Huang J, Sheehy J, Laza R, Visperas R, Zhong X, Centeno G, Khush G, Cassman K (2004) Rice yields decline with higher night temperature from global warming. Proc. Natl. Acad. Sci. USA 101: 9971–9975.

Population Reference Bureau (2007) The future population of India. A long range demographic view. http://www.prb.org/pdf07/futurepopulationofindia.pdf (accessed on August 20, 2013).

Pradhan GP, Prasad PVV, Fritz AK, Kirkham MB, Gill BS (2012a) Effect of drought and high temperature stress on synthetic hexaploid wheat. Funct. Plant Biol. 39: 190–198.

Pradhan GP, Prasad PVV, Fritz AK, Kirkham MB, Gill BS (2012b) High temperature tolerance in *Aegilops* species and its potential transfer to wheat. Crop Sci. 52: 292–304.

Prasad PVV, Djanaguiraman M (2011) High night temperature decreases leaf photosynthesis and pollen function in grain sorghum. Funct. Plant Biol. 38: 993–1003.

Prasad PVV, Craufurd PQ, Summerfield RJ (1999a) Sensitivity of peanut to timing of heat stress during reproductive development. Crop Sci. 39: 1352–1357.

Prasad PVV, Craufurd PQ, Summerfield RJ (1999b) Fruit number in relation to pollen production and viability in groundnut exposed to short episodes of heat stress. Ann. Bot. 84: 381–386.

Prasad PVV, Craufurd PQ, Summerfield RJ (2000a) Effect of high air and soil temperature on dry matter production, pod yield and yield components of groundnut. Plant Soil 222: 231–239.

Prasad PVV, Craufurd PQ, Summerfield RJ, Wheeler TR (2000b) Effects of short episodes of heat stress on flower production and fruit-set of groundnut (*Arachis hypogaea* L.). J. Exp. Bot. 51: 777–784.

Prasad PVV, Boote KJ, Allen Jr LH, Thomas JMG (2002) Effects of elevated temperature and carbon dioxide on seed-set and yield of kidney bean (*Phaseolus vulgaris* L.). Glob. Change Biol. 8: 710–721.

Prasad PVV, Boote KJ, Allen Jr LH, Thomas JMG (2003) Super-optimal temperatures are detrimental to reproductive processes and yield of peanut under both ambient and elevated carbon dioxide. Glob. Change Biol. 9: 1775–1787.

Prasad PVV, Boote KJ, Allen Jr LH (2006a) Adverse high temperature effects on pollen viability, seed-set, seed yield and harvest index of grain sorghum [*Sorghum bicolor* (L.) Moench] are more severe at elevated carbon dioxide due to higher tissue temperatures. Agri. For. Meteorol. 139: 237–251.

Prasad PVV, Boote KJ, Allen Jr LH., Sheehy JE, Thomas JMG (2006b) Species, ecotype and cultivar differences in spikelet fertility and harvest index of rice in response to high temperature stress. Field Crop Res. 95: 398–411.

Prasad PVV, Pisipati SR, Mutava RN, Tuinstra MR (2008a) Sensitivity of grain sorghum to high temperature stress during reproductive development. Crop Sci. 48: 1911–1917.

Prasad PVV, Pisipati SR, Ristic Z, Bukovnik U, Fritz A (2008b) Impact of high nighttime temperature on growth and yield of spring wheat. Crop Sci. 48: 2372–2380.

Prasad PVV, Pisipati SR, Momcilovic I, Ristic Z (2011) Independent and combined effects of high temperature and drought stress during grain filling on plant yield and chloroplast EF-Tu expression in spring wheat. J. Agron. Crop Sci. 197: 430–441.

Redden RJ, Yadav SS, Hatfield JL, Prasanna BM, Vasal SK, Lafarge T (2011) The potential of climate change adjustment in crops: A synthesis. In: Yadav SS, Redden RJ, Hatfield JL, Lotze-Campen H, Hall AE (eds). Crop adaptation to climate change. Wiley-Blackwell, Chichester, pp. 482–494.

Rekika D, Monneveux P, Havaux M (1997) The in vivo tolerance of photosynthetic membranes to high and low temperatures in cultivated and wild wheats of the *Triticum* and *Aegilops* genera. J. Plant Physiol. 6: 734–738.

Ristic Z, Bukovnik U, Momcilovic I, Fu J, Prasad PVV (2008) Heat induced accumulation of chloroplast protein elongation factor, EF-Tu in winter wheat. J. Plant Physiol. 165: 192–202.

Sacks WJ, Kucharik CJ (2011) Crop management and phenology trends in the U.S. corn belt: Impacts on yields, evapotranspiration and energy balance. Agri. For. Meteorol. 151: 882–894.

Saini HS, Sedgley M, Aspinall D (1983) Effect of heat stress during floral development on pollen tube growth and ovary anatomy in wheat (*Triticum aestivum* L.). Aust. J. Plant Physiol. 10: 137–144.

Scafaro AP, Haynes PA, Atwell BJ (2010) Physiological and molecular changes in *Oryza meridionalis* Ng., a heat-tolerant species of wild rice. J. Exp. Bot. 61: 191–202.

Schlenker W, Roberts MJ (2009) Nonlinear temperature effects indicate severe damages to U.S. crop yields under climate change. Proc. Nat. Acad. Sci. USA 106: 15594–15598.

SPIA Draft Report (2011) Global and regional trends in production, trade and consumption of food legumes. 45pp. http://impact.cgiar.org/sites/default/files/images/Legumetrends.pdf (accessed on August 20, 2013).

Tanner CB (1963) Plant temperature. Agron. J. 55: 210–211.

Tata Strategic Management Group (2012) Pulses: A world view. http://www.growmorepulses.com/article.aspx?cont_id=0Xw9Kk6LW6s= (accessed on August 20, 2013).

Tausz M, Tausz-Posch S, Norton RM, Fitzgerald GJ, Nicolas ME, Seneweera S (2013) Understanding crop physiology to select breeding targets and improve crop management under increasing atmospheric CO_2 concentrations. Environ. Exp. Bot. 88: 71–80.

Waines JG (1994) High temperature in wild wheats and spring wheats. Aust. J. Plant Physiol. 21: 705–715.

Wang YP, Connor DJ (1996) Simulation of optimal development for spring wheat at two locations in southern Australia under present and changed climate conditions. Agri. For. Meteorol. 79: 9–28.

Wassmann R, Nelson GC, Peng SB, Sumfleth K, Jagadish SVK, Hosen Y, Rosegrant MW (2010) Rice and global climate change. In: Pandey S, Byerlee D, Dawe D, Dobermann A, Mohanty S, Rozelle S, Hardy B (eds). Rice in the global economy: Strategic research and policy issues for food security. International Rice Research Institute, Los Banos.

Welch JR, Vincent JR, Auffhammer M, Moya PF, Dobermann A, Dawe D (2010) Rice yields in tropical/subtropical Asia exhibit large but opposing sensitivities to minimum and maximum temperatures. Proc. Natl. Acad. Sci. USA 107: 14562–14567.

Wellinger A (2007) Algae biomass. Does it save the world? Short reflections. IEA Bioenergy. http://www.iea-biogas.net/_download/publi-task37/AlgaeBiomass8-09.pdf (accessed on August 20, 2013).

Wheeler TR, Craufurd PQ, Ellis RH, Porter JR, Prasad PVV (2000) Temperature variability and the yield of annual crops. Agric. Ecosyst. Environ. 82: 159–167.

Wiegand CL, Namken LN (1966) Influences of plant moisture stress, solar radiation, and air temperature on cotton leaf temperature. Agron. J. 58: 582–586.

Index

Note: Italicized page locators indicate photos/figures.

Temperature and Plant Development, First Edition. Edited by Keara A. Franklin and Philip A. Wigge.